what
every
science
student
should
know

Chicago Guides to Academic Life

what every science student should know

Justin L. Bauer, Yoo Jung Kim,
Andrew H. Zureick, and
Daniel K. Lee

The University of Chicago Press
Chicago and London

Justin L. Bauer is a medical student at the University of California, San Diego. **Yoo Jung Kim** is a medical student at Stanford University. She served as editor-in-chief of the *Dartmouth Undergraduate Journal of Science*. **Andrew H. Zureick** is a medical student at the University of Michigan. He served as editor-in-chief and president of the *Dartmouth Undergraduate Journal of Science*. **Daniel K. Lee** is a medical student at Harvard Medical School. He served as editor-in-chief and president of the *Dartmouth Undergraduate Journal of Science*.

The University of Chicago Press, Chicago 60637
The University of Chicago Press, Ltd., London
© 2016 by Justin L. Bauer, Yoo Jung Kim, Andrew H.
Zureick, and Daniel K. Lee
All rights reserved. Published 2016.
Printed in the United States of America

25 24 23 22 21 20 19 18 17 16   1 2 3 4 5

ISBN-13: 978-0-226-19874-3 (cloth)
ISBN-13: 978-0-226-19888-0 (paper)
ISBN-13: 978-0-226-19891-0 (e-book)
DOI: 10.7208/chicago/9780226198910.001.0001

Library of Congress Cataloging-in-Publication Data

Names: Bauer, Justin L. (Justin Lawrence), 1990–author. | Kim, Yoo Jung, 1991–author. |
    Zureick, Andrew H. (Andrew Harrison), 1991–author. | Lee, Daniel K. (Daniel Kwang-
    suk), 1991–author.
Title: What every science student should know / Justin L. Bauer, Yoo Jung Kim,
    Andrew H. Zureick, and Daniel K. Lee.
Other titles: Chicago guides to academic life.
Description: Chicago : The University of Chicago Press, 2016. | Series: Chicago guides to
    academic life | Includes bibliographical references and index.
Identifiers: LCCN 2015041270| ISBN 9780226198743 (cloth : alk. paper) | ISBN
    9780226198880 (pbk. : alk. paper) | ISBN 9780226198910 (e-book)
Subjects: LCSH: Science–Study and teaching (Graduate)–United States. | Universities and
    colleges–United States–Graduate work. | Academic achievement–United States.
Classification: LCC Q181.A2 B38 2016 | DDC 507.1/173–dc23 LC record available at http://
    lccn.loc.gov/2015041270

♾ This paper meets the requirements of ANSI/NISO
Z39.48–1992 (Permanence of Paper).

To our parents, William Bauer and Nancy Spiegel, Hyea Ja Chae and Won Sik Kim, Samir and Brenda Zureick, and Daniel and Linda Lee, for their love and support

Contents

Acknowledgments

This book would not exist had it not been for the help of people who believed in us. First, we thank Andrea Somberg of Harvey Klinger Inc. for her guidance in helping us to formulate the initial book proposal. Next, we thank Christie Henry, Amy Krynak, Logan Ryan Smith, and Lauren Salas at the University of Chicago Press for guiding us through the publishing process and Mary Corrado for her editorial acumen.

We thank all of the students, faculty, professionals, and young college alumni who generously shared with us their time and insights about college science through interviews and reviews of our manuscript. Our particular thanks go out to Joshua Alman, Delian Asparouhov, Nancy Bakowski, Carli Balogh, Ameen Barghi, William D. Bauer, PhD, Mark Baum, Jacob Becraft, Sydney Behrmann, Alysia Birkholz, PhD, Scott Brookes, Kristopher S. Brown, Michael Cariaso, Hongyu Chen, Jeff Chen, Kenny Chen, Sara Choi, Cesar Cuenca, Irving Dai, Ellen Daily, Donna J. Dean, PhD, Amar Dhand, MD, DPhil, Christiane Donahue, PhD, Jaideep Dudani, Kar Epker, Riley Ennis, Roxanne Farkas, Vanessa Ferrel, Christopher Finch, Emily Flynn, Sydney Foote, Christopher Francis, MS, Mollie Sarah Henni Friedlander, Ryan Gabelman, Amanda Gartside, Angela Gauthier, Sophia Gauthier, MS, Marcelo Gleiser, PhD, David S. Glueck, PhD, Charles Goldberg, MD, Samuel Greene, MS, Natasha M. Grotz, PhD, Julie Ann Haldeman, Michael Huarng, Colin Heffernan, Steven Jin, Carlee Joe-Wong, Stephan Johnson, Peter Kalugin, Julia E. Kao, Christopher S. Kelly, Roger Khouri Jr., Yoo Eun Kim, Clarke Knight, Aaron Koenig, F. Jon Kull, PhD, Nilay Kumar, Daniel Sotelo Leon, Jonathan Li, Grant L. Lin, George S. Liu, William Lotko, PhD, Jan A. Makkinje, Rachel Mann, Daniel Marcusa, Vicki V. May, PhD, Sara Koenig McLaughlin, PhD, Kelvin Mei, Roland Nadler, JD, Aran Nayebi, Nicole Nevarez, Paloma Marin Nevarez, Stephen Neville, Ben H. Nguyen, David Nykin, Matthew L. Pleatman, Mya Poe, PhD, Robert Porter, Gareth Roberg-Clark, Juan Pablo Ruiz, Parker Phinney, Rameshwar R. Rao, Raza Rasheed, JD, Stanford Schor, Anthony Scruse, Elisabeth Seyferth, David Shafer, Eric Shen, Maxwell Shinn, Alvin Siu, Jessica Smolin, PhD, Leslie J. Sonder, PhD, Nancy H. Spiegel, MS, Alison Stace-Naughton, Alyssa Stevenson, Mika M. Tabata, Michael Terjimanian, Kaya Thomas, Carl P. Thum, PhD, Jonathan D. Tijerina, MS, Duy C. Tran, Jeffrey Treiber, Jeffrey Tsao, Ryan Tsuchida, Christopher Walker, Sara Walker, Kathy S. Weaver, MA, Elise Wilkes, William Wingard, Lee A. Witters, MD, Xiaotian Wu, Jennifer Xia, So Young Yang, Joseph K. Yi, Lindsey Youngquist, Melanie Zhang, and Yingchao Zhong.

Lastly, we acknowledge our friends and faculty members from Dartmouth College for helping to instill within us our love for the sciences. Our gratitude also goes out to the other institutions and organizations that helped us to work on this book, such as the Dartmouth College Institute for Writing and Rhetoric, Dartmouth College Office of Undergraduate Research, Dartmouth Undergraduate Journal of Science, and the Barry Goldwater Scholarship and Excellence in Education Program.

1 Welcome to the World of College Science

The one thing none of your college science courses will teach you is how to succeed in them. Studying science needs to come with an owner's manual, and that manual is this book.

Study skills, choosing a major, research, and career planning are just a few of the topics covered in this concise guide. The scientific disciplines—math, engineering, chemistry, computer science, etc.—are both challenging and rewarding. Yet relatively few students make it through the intense and sometimes competitive world of college science. Because you don't know what you don't know, you need advice from people who have been through what you are about to experience. This book—painstakingly distilled from years of research, interviews with successful scientists and science students, and our own experiences as recent science graduates—is the advice that we, your authors, wish we had heard when we came to college.

Good luck and welcome to the world of college science! We look forward to helping you every step of the way in the coming pages.

Why Is Science So Hard?

Only a small fraction of the most able youngsters enter scientific careers. I am often amazed at how much more capability and enthusiasm for science there is among elementary school youngsters than among college students. Something happens in the school years to discourage their interest (and it is not mainly puberty); we must understand and circumvent this dangerous discouragement. No one can predict where the future leaders of science will come from.

Carl Sagan[1]

Sixty percent of college students planning to study science or medicine change their minds later in their academic careers.[2] Why do so many students end up leaving their scientific aspirations behind?

First of all, science isn't easy to learn. We come out of the womb with the capacity to learn human language, but no one begins life with the instinctive ability to understand quantum physics. To learn science, you have to work at it, like most things worth doing. But this isn't the full story; in fact, the natural challenge of learning difficult concepts isn't the biggest reason students struggle with science. Rather, many college students are dissuaded from science because they don't know *how* to prepare for their college science courses.

Science classes—sometimes also called STEM classes for Science, Technology, Engineering, and Mathematics—can be complicated, impersonal, and often confusing (Note: we will use the terms *science* and *STEM* interchangeably). In a typical introductory college science lecture, you will find yourself in a huge room with hundreds of other students, straining to focus on the tiny professor at the front of the class as she whips through her presentation or draws complicated equations on the board. Some students get it right off the bat, and some students don't. Oftentimes, at the end of the course, all you are given is a test and a grade. If the grade is too low, many students simply call it quits.

Don't Give Up!

While there are a number of roadblocks that discourage students from pursuing science, there are even more reasons to stick with it. With lots of hard work and the guidance of this book, you will be able to excel in your classes, earn your college degree, and, perhaps most importantly, appreciate the beauty of the science you study.

But science is not just interesting to study. There are also very practical benefits to earning a STEM degree. The professional world needs and is prepared to pay for people with skills in science and mathematics. Over the past several decades, the percentage of students graduating from college with a STEM degree has declined, while the demand for science-related jobs has grown and will continue to grow.[3] From 2001 to 2011, growth in STEM jobs was three times faster than growth in non-STEM jobs.[4] When STEM majors graduate, on average, they make more than professionals with other degrees.[5] One study found that science majors would earn half a million dollars more than other majors over the course of their lifetime.[6]

The professional opportunities that a science major provides are not limited to science-related fields. In fact, STEM graduates have higher salaries than other majors, regardless of whether or not they work in a STEM-related occupation.[7] Take a guess, what college major is most common amongst S&P 500 CEOs? Business? Economics? Marketing? No. Actually, it's engineering.[8] This just goes to show that the skills you can learn in college as a science major are prized in a wide range of fields. According to the National Science Foundation[9] and the

Department of Labor,[10] 80% of *all* the jobs created in the next decade will require math and science skills. Getting these skills is smart. When you graduate, quite possibly with significant debt, you'll be happy to know that you've amassed valuable and marketable skills to begin making your way in the world.

Finally, from a much broader perspective, skilled scientists are crucial to our future. Modern science touches every aspect of our lives from the produce in your local grocery store to life-saving pharmaceuticals to the safety features of your car. The responsibility to meet the biggest problems of our century will rest on the shoulders of our scientists—challenges like curing diseases and finding clean energy resources. The prospect of helping the world address such challenges is yet another incentive for those considering studying science.

Studying for Skills, not Just Grades

As a college student, you'll need to approach your academic life with a whole different attitude than you had in high school. Many aspiring science students feel crushed when they get a bad test score, perhaps for the first time in their lives. They think a low grade means they are bad at science. This misconception is one of the biggest reasons that students give up on STEM. But keep this in mind: doing badly in a science course *doesn't* necessarily mean you are bad at science or a bad student.

At Dartmouth College, what do a theater major with a 3.89 GPA and a chemistry major with a 3.11 GPA have in common? Their grades are both equal to the average grade given by courses in their respective departments.[11] Across the board, arts, social sciences, and humanities courses give out higher grades than science courses.[12] Your grade may be more of a reflection of a departmental policy or a quota set by an instructor than of your true talents and interests. Grades are important, but they certainly aren't everything, and they may even be misleading.

In high school, everybody studied more or less the same subjects, so the main factor that differentiated students academically was their GPA. In college, students study different subjects, so comparing grades between majors is like comparing apples and oranges. If you don't have a trust fund in your name (or even if you do), it will be important for you to graduate with knowledge and skills you can apply in the working world, regardless of what grades you get.

In *Forbes* magazine's ranking of the ten college majors with the worst employment prospects and the worst salary after graduation, all ten of the worst majors were nonscience majors.[13] The college graduates who received high grades in those majors probably don't care very much anymore about how well they were doing on paper. High school was about getting good grades. In college, you need to be studying for grades *and* skills.

Getting Started

We started writing this book as college students because we saw our classmates in the sciences dwindle in number year after year and were stunned to find out that this was a nationwide phenomenon that no one had successfully addressed. After three years of research, interviews, and writing, we put together a book compiled from the advice of students and recent graduates who have excelled academically, presented in national symposiums, published in journals, created apps, and started their own businesses—all while earning their bachelor's degrees. To make the book as relevant as possible to the average student, we interviewed students from a wide variety of STEM majors from small liberal arts schools, to research-focused private schools, to major public universities, and everything in between. Much of the advice in this book comes from recipients of high academic honors like the Barry Goldwater Scholarship, Fulbright Study/Research Grant, Churchill Scholarship, Gates Cambridge Scholarship, Marshall Scholarship, and the Rhodes Scholarship.

Whether you are a college student still navigating the lay of the land or an ambitious high school student looking for a head start, this book will provide you with the basic knowledge to tackle science head-on and excel in college and beyond.

Chapter-by-Chapter Overview

We've included a brief synopsis of each chapter below. Each chapter can be read and understood on its own, but there is a logical progression from one chapter to the next. Even if you think you already know about the topic we discuss in a given chapter, we still encourage you to read it. Sometimes it is what we *think* we know already, that makes it hardest for us to learn. As Mark Twain once said: "It ain't what you don't know that gets you into trouble. It's what you know for sure that just ain't so."

Chapter 2: How to Manage College Life

College is an exciting time with seemingly endless opportunities. However, if you aren't cautious, this can be to the detriment of the your grades, as new college students often fail to balance their personal and social lives with their academic responsibilities. In this chapter, we talk about how to manage time and thereby develop the foundation to lead an efficient and satisfying college life.

Chapter 3: How to Excel in Your STEM Courses

This chapter will help you hone your academic skills to succeed in the classroom and in laboratory courses. We discuss tips for taking notes, reading textbooks, preparing for quizzes and exams, and writing laboratory reports.

Chapter 4: Choosing a STEM Major

Here, we will introduce you to the most general types of science majors, explain what those majors are like, and acquaint you with the types of careers that each major tends to pursue.

Chapter 5: Conducting Scientific Research

Undergraduate research is an essential experience for students interested in a career in science, be it academic, medical, or industrial. Through this chapter, you can familiarize yourself with the world of academic research and ins and outs of conducting a research project as a student. Additionally, you'll become acquainted with some of the unique vocabulary, hierarchy, and unwritten rules of the research culture.

Chapter 6: Beyond Your Bachelor's Degree

How do you go from being a student to being a professional? This chapter will help you make this transition by giving you tools for finding a job after college, such as writing your personal statement, creating your portfolio, and applying for scholarships and fellowships.

Chapter 7: STEM in the Real World
This chapter provides detailed advice about preparing for graduate school and professional schools (e.g., medical, law, and business school), and general information about careers for students with a background in science.

Chapter 8: In Conclusion
We close this book with several important tips for all college students, especially STEM students, for rounding out your undergraduate academic experience and preparing for what lies ahead.

Appendix: Advice for Underrepresented Students in STEM
This chapter explains some of the challenges faced by women and other underrepresented students in the sciences as well as some of the opportunities available to these groups. Additionally, we discuss the importance of mentorship in a successful academic experience.

A note about gender pronouns: we shift back and forth between the pronouns "he" and "she" in this book when referring to students, professors, counselors, etc. Science is for everyone, but the English language often makes it awkward to form a sentence without specifying a gender pronoun.

Buckle Up, and Enjoy the Ride!
Studying STEM in college will be an incredible experience, but it will also require a tremendous amount of work and dedication. Fortunately, you've already taken an enormous step by picking up this book and reading this far. That alone speaks volumes about your dedication to succeed in college. We look forward to helping you channel that commitment into the improvement of your college STEM experience. Let's begin.

2 How to Manage College Life

Time is what we want most, but what we use worst.
William Penn

The typical student arrives on campus excited and nervous about the coming year. He moves into his dorm room, signs up for interesting courses, makes new friends, and starts receiving invitations to join all sorts of extracurricular activities from mock trial to rock climbing. As the term goes on, he gets progressively more exhausted. He starts missing classes. His laundry starts piling up. A fuzzy black mold begins to grow on his shower curtains, and—to his horror—he doesn't know how to clean it. He feels as if he is spread too thin with classes, club events, and social obligations demanding his time and attention, but he is reluctant to drop any of his commitments due to a fear of missing out on novel experiences. Suddenly, at the end of the term, finals period rears its ugly head, and he focuses for days in a state of caffeine-fueled jitteriness only to find out that he scored much more poorly than he ever did in high school. Only then does he realize that he needs a new strategy.

College demands a different attitude toward work and play than what may have worked for you in high school. Academic expectations are higher, and you have more control over your life than ever before. This chapter and the next will help you adjust to a new pace of working by teaching you to manage your college life and prepare to excel in your STEM coursework. If you take these lessons to heart, you can avoid repeating this frequent college pattern by anticipating the challenges that you will face in college and planning accordingly.

Success: It's Personal

Comparison is the thief of joy.

Theodore Roosevelt

Before you set out to "succeed" in college, you need to define what that means to you. Is it earning good grades, conducting meaningful research, gearing up for a specific profession, or something else entirely?

Your idea of success may not be the same as that of your friends, your professors, or your parents. Take some time for self-reflection, sort out what is most important to you, and think seriously about what you would like to get out of college and in your career. One University of Michigan student whom we interviewed summed this thought up nicely: "Always proceed with the end in mind. Yes, you have time to figure out what you enjoy and want to go into, but it should bother you if you do not know where you expect to be after graduation. If you don't have a plan, devote your spare time to figuring it out. It isn't the smartest people who succeed in life, but the most driven." The sooner you figure out what your own goals are, the better you can decide how to get there and what you need to do during college.

You and your classmates come from a wide range of academic backgrounds—from inner-city public schools to prestigious preparatory programs. At the beginning of school, students will find themselves at different levels of readiness to tackle college work. In general, however, your college classmates will be more intelligent and driven than your high school peers; after all, that's how they got to college in the first place. Moreover, your classmates will have different amounts of responsibilities in their lives. Some students may be earning money through work-study to pay for their education; others may be preoccupied with family obligations; still others will be free to devote their whole energy and attention to their classes.

Define success in college in a way that takes into account your educational background and responsibilities. Continually set short- and long-term goals that will leave you satisfied at the end of the day, term, and year. A goal is both a destination and an anchor: you have to chart out what success means for you before you can get there, and having your objective in mind will keep you from going adrift.

Regardless of where you place yourself in the spectrum of college preparedness, don't let your assessment of your peers' academic abilities lead you to doubt your own abilities or lure you into thinking that college is going to be a breeze. More so than anything else, your performance in class depends on your willingness to work hard. If you find yourself having less science background than some of your classmates, it may take some extra time and effort, but by keeping at it, you will catch up.

When you've decided what your goal is, write it down and post it in a location that will be visible to you: on top of your desk, on the door of your room, across from the toilet, wherever. Every time you feel discouraged, remind yourself of what you are working for and hold yourself accountable for realizing your goals. But don't psyche yourself into thinking that you have to stick to the first goal that you've set for yourself. College is a period for development, and you may realize what you wanted to accomplish more than anything as a freshman is no longer relevant for you during senior year. Be open to changing your destination, but always think about how you can get there.

Students Say: How Can Students Find Their Own Meaning of Success?

Know yourself and what you want to get out of your education. You can always make time for something if it is important to you. Also, know your limits, as it is always better to do a few things well than to do many things half-heartedly.

Max, University of Minnesota, Goldwater Scholar, Churchill Scholar

The most important part of being successful is finding something that really excites you and figuring out how to focus on that throughout your career. Explore as much as you can during your first few years of college and think about which aspects of what you explore grab you the most.

Sam, University of Chicago, Goldwater Scholar, Rhodes Scholar

Time management and your own definition of success in college accompany one another. If athletics and academics both matter to you, as they did for me, you will find a way to do both. What may seem like sacrifices to others, such as reduced social time, will be easy decisions for you if you have a strong, self-motivated sense of what matters. Developing this sense requires introspection. You can learn much about yourself by considering your own thoughts and embracing challenges. Do not be afraid to think for yourself and do not shy away from courses in other disciplines. You will only become more independent, resilient, and creative.

Chris, Amherst College, Goldwater Scholar, Churchill Scholar

For most of my life, I've defined success by how much I was living up to others' (namely, [my] parents') expectations, whether it was getting good grades or getting into a top whatever college. But having the freedom to do what I believed in and love in college has taught me to rethink my definition of success. Because no matter how happy other people are for you about getting into a prestigious school or taking on a crazy course load, if you are absolutely miserable and forced into doing something you don't like, then what's the point of being "successful" in other people's eyes? This should be obvious, but the best scientists are not those who do it for the degree, money, or prestige, but rather the people who are truly passionate about the discovery-making process and contributing to the ever-growing world of scientific knowledge.

Melanie, Emory University, Goldwater Scholar

The Importance of Time Management

Time management is really a misnomer—the challenge is not to manage time, but to manage ourselves.

Stephen Covey

College will probably be the first time in your life that you have almost complete autonomy, with no one to tell you why eating microwavable burritos for breakfast, lunch, and dinner is a terrible idea. Such liberty can be exciting, but with all the new social and extracurricular opportunities of college demanding your attention, you may find it difficult to keep focused on your classes and your well-being. Because of this, the first step to transitioning into college is learning *how* to be independent, and a key part of that is learning how to manage your time.

In college, time spent in class is only a fraction of what it was in high school—perhaps as little as an hour or two in a given day. Attendance is rarely taken, especially in huge introductory lecture courses. Given this lack of supervision, it's up to you to build some structure into your life to balance your academic, social, and extracurricular tasks.

The following section discusses time management methods that have worked well in our own experience and for the many successful STEM students whom we've interviewed. If you apply these techniques diligently, you will be well on your way to getting the most from your college experience. It will require discipline to establish this practice, but it gets easier with time and will be well worth it in the long run.

Every Term: Use a Planner

Get a planner and fill it with your short- and long-term goals and course assignments. As soon as you get your hands on your class syllabi at the beginning of each term, take the time to analyze them and jot down due dates for lab reports, projects, and tests for all of your classes, in addition to any important social obligations and fun events (e.g., birthdays, concerts, parties).

By understanding your schedule early in the term, you can pinpoint your "Hell Weeks," those unavoidable stretches when you have to juggle multiple papers, tests, and other commitments. This way, you'll know when you have to put in more work than usual and you can get yourself prepped for it in advance. Finally, list any personal or academic goals you have for yourself and make the time to achieve them. For instance, if you want to ace a particular course, set aside an extra chunk of your time before every major test and assignment to ensure that you will be able to focus on that class.

Physical planners and digital calendars are both good options to keep organized. A digital calendar like iCal or Google Calendar will let you set reminders before events (e.g., desktop notifications or auto-generated text messages). Do what works best for you and stick to it; planners will keep you organized, prompt, and focused on your goals.

Keep Track of Your Time

In addition to filling out your planner for the academic term, plan your schedule on a weekly basis. Ask yourself how you can use each week to keep up with your classes, lead a healthy lifestyle, and still have time for fun. Take fifteen minutes every weekend and write down your academic, social, and personal goals for the coming week into your planner. How many labs, problem sets, and readings do you have this week? Would after class on Tuesday be most convenient to get lunch with your roommate? Do you need to do your laundry this week or can that wait until next week? Plan out your tasks every week, reviewing them each night or early morning to greet the new day with a sense of direction. Soon this type of thinking will become second nature.

A weekday schedule might look something like the following:

Today's TO DO List

8:00 to 8:45 a.m.: Wake up and eat breakfast

8:45 to 9:45 a.m.: Preview and prepare for class

10:00 to 11:00 a.m.: Lecture #1

11:10 a.m. to 12:10 p.m.: Lecture #2

12:15 to 1:00 p.m.: Lunch and break #1

1:10 to 3:10 p.m.: Review lectures and solve practice problems

3:20 to 4:20 p.m.: Hit the gym

4:30 to 5:30 p.m.: Go grocery shopping

5:40 to 6:40 p.m.: Do laundry and review flashcards

6:50 to 7:50 p.m.: Dinner and break #2

8:00 to 8:50 p.m.: Attend club meeting

9:00 to 10:30 p.m.: Do practice problems and problem sets

10:30 to 11:30 p.m.: Break #3

11:30 p.m. Get ready for bed

Take note of two things in this schedule. First of all, much of the day is accounted for and there are buffer times between activities to account for time needed to get from place to place. Secondly, notice that free time is built right into the schedule; this way, you'll be able to enjoy your breaks without feeling guilty or going overboard and shoving aside all the work you still have left. This schedule is comprehensive and covers the time that you spend in class, studying, sleep, meals, exercise, errands, and extracurricular activities. Careful planning will help you be focused when you need to work and relaxed when you need to rest.

That being said, you will almost certainly deviate from your schedule. Your problem set may be harder to solve than you'd expected, a quick dinner with friends may turn into a lengthy conversation, or you may just feel like you need a break. Leave some flexibility for the unexpected. At the end of each day, review your schedule, cross off everything you've accomplished, and reschedule anything you couldn't finish for later. Figure out what kept you from getting things done (not having enough time, procrastination, etc.) and think about how you can avoid these pitfalls next time. Keep making your schedules for each day even if you don't end up following them to the letter; writing down your schedule is an exercise in self-discipline and planfulness, not an attempt to tell the future.

Divide and Conquer Any Task

Whether it's studying for a final, writing a paper, or planning the party of the century, simplify any task into a checklist of smaller mini-tasks and jot it in your planner or on a note card. Make each mini-task as simple as possible and set a deadline for each item on your checklist. The process of making the checklist will make you think seriously about how you can approach a given problem and will make the task—no matter how big—seem less overwhelming. The deadline for each mini-task will keep you on track and help you fit them in your daily calendar. Carry the list with you to remind yourself what still has to be done.

There's also something deeply satisfying about checking off each mini-task as it is completed, but don't just take our word for it! Here are a few examples of how you can take a complicated task and break it into measurable bite-sized pieces:

Example Task #1: Organize a club dinner to raise funds for a charity
- Get catering quotes from food vendors (10/5)
- Book venue from Facilities Department (10/6)
- Finalize programming for the dinner (10/9)
- Design and print out flyers (10/10)
- Send out email invitation to the campus and ask for RSVPs (10/11)
- Update food vendor with estimated food orders (10/18)
- Set up and run the event (10/23)

Example Task #2: Study for the final
- Review notes and briefly skim through the text (12/15)
- Review previous problem sets and midterm (12/16)

- Go to office hours to ask about questions that I haven't been able to figure out (12/16)
- Take practice final 1 and review answers (12/17)
- Take practice final 2 and review answers (12/18)
- Go to office hours to ask about questions that I haven't been able to figure out (12/18)
- Go over concepts that I still feel weak on (12/19)
- Take the test (12/20)

Example Task #3: Major research paper
- Pick a topic (1/20)
- Read 5 to 8 scholarly papers supporting the topic and keep a bibliography (1/21)
- Create comprehensive outline (1/22)
- Flesh out the outline (1/23)
- Finish first draft of paper (1/24)
- Major revision #1 (1/25)
- Major revision #2 (1/26)
- Ask friends for constructive feedback (1/26)
- Break (1/27)
- Major revision #3 + grammar check (1/28)
- Turn in the paper (1/29)

Optimize Your Work Time

Consider this scenario: It's two hours before a quiz, and you haven't prepared for it at all. What do you do? Most likely, you crack open your notes and textbook, and you work intensely to try to cram as much information into your head as possible. With the clock biting at your heels, you absorb more in those two hours than you would in four or even five hours of your regular study sessions. Although cramming should *never* be a part of your planned work regimen, this is the level of focused intensity you should aim for every time you study. Why should you spend four or five hours doing something you can do just as well in two?

When you sit down to study, squeeze as much work out of your time as you possibly can. This could involve working alone, going to the library, studying in groups, whatever it is that allows you to best concentrate on the material. If you don't know what sort of setting will help you to be more productive, rotate through different study environments to see which one fits your learning style (more on this on chapter 3).

Many students like to listen to music or reruns of their favorite TV shows playing in the background when they study. They say this helps them focus, but to be honest, these are all distractions that bleed your concentration. Time and time again, research has shown that people can't multitask as well as they think they can and are only really capable of doing one thing at time.[1] Good, efficient studying means you should be able to have complete focus on the material. In fact, for many students, any noise gets in the way of achieving optimal concentration. This is why some students use earplugs or other noise-cancelling devices. Try it out. You'll be surprised at the amount of ambient noise you block out. There is a reason why many standardized test centers allow test takers to wear earplugs. Students want to block out the noise of the other test takers because this helps them put their absolute focus into the test. Why not do the same every time you study?

A final method for maximizing your study efficiency is to time yourself while studying. Set a timer for 50 minutes right before you start your work and promise yourself that as long as the timer is running you will do nothing other than study.

When the timer goes off, take a 10-minute break to go to the bathroom, get some water, or check your phone, and start preparing for the next 50-minute block of work. A solid, efficient hour of focus is preferable to three hours of problem solving while simultaneously checking email, using social media, texting, and watching videos. The timer will tell you to drop everything and work.

Don't Be Afraid to Say "No"

One of the best things about college is the amazing variety of fun things to do. There are clubs and events to satisfy the most esoteric interests from soufflé baking to Indian classical music. Moreover, many of your new friends will be close by to do these things with you. You will see them in class, at the dining hall, and, if you live in a dorm, at home.

With so many exciting events and interesting people surrounding you in college, deciding how to spend your time can be an arduous task. There are only so many hours in a day, so you will need to prioritize between what you want to do and what you need to do. Your friends may come knocking on your door to invite you to a party at 11 PM. You may want to polish your fiery oratory skills on the college debate team. A professor may offer you a position as a teaching assistant (TA) in her class. In all these cases, you need to think carefully about whether you have the time for it. If you don't, you need to be comfortable saying no, both to the person who has requested your time and attention and to yourself.

In our ever-connected lives, it's hard turning down an interesting opportunity due to a Fear of Missing Out (FOMO), defined by the *Oxford English Dictionary* as "anxiety that an exciting or interesting event may currently be happening else-

where, often aroused by posts seen on a social media website."[2]
Because of FOMO, many college students stretch themselves
too thin with clubs and commitments—becoming a jack-of-
all-trades and a master of none. Treat your time like the pre-
cious resource that it is. Guard it carefully and spend it on
the people and goals that matter most to you. It's okay to try
new things in college—in fact, college is the perfect time for
that! But remember that you are first and foremost a student,
so don't feel left out if you have to say no to people in order to
focus on your academics or on yourself. If you only have time
for your schoolwork, your friends and family, and one or two
extracurricular activities, this is perfectly fine.

Summary
- Keep track of your time
- Divide and conquer any task
- Get the most of your work time through sustained focus
- Don't be afraid to say "no"

Use the four general points we've described above to be
organized and efficient every day. This advice will ensure that
you never miss a deadline. If you find yourself not accomplish-
ing as much as you hope to, speak with an academic advisor,
such as an undergraduate dean, an academically successful
upperclassman, or your professors. You may need to re-allocate
your time between your social, academic, and extracurricular
commitments. Maybe you will find that you need to improve
your study methods (a topic that will be explored in chapter
3). Above all, stick to your schedule and avoid putting off tasks
that you can do right now.

We can't emphasize enough how crucial it is to learn time

management as you start your college career. Successfully managing your time will enable you to maximize your productivity and imbue all your actions with a sense of purpose.

Students Say: How Did You Make the Most of Your Time?

The transition from a small town in East Texas to an urban academic environment was a bumpy one for me. It simply took a while to sink in that at a certain level one cannot play so many sports while maintaining good academic standing and fulfilling involvement in extracurricular activities. Specialization drives our world at the present, and I encourage you to keep in mind that if you spread yourself too thin, you risk jeopardizing the quality of each of your individual pursuits. So choose carefully what to invest in. Dedicate yourself to it and you will enrich those things most important and avoid being overwhelmed.

Jonathan, Baylor University

There are never enough hours in the day—and that's perfectly OK. What's important is to use the time you have efficiently because every minute counts. From biking instead of walking to going over those lecture notes while waiting for that gel to finish running in lab, the little things start to add up. There were days I'd spend more time in lab than in class, and it wasn't a huge deal because it helped me learn to adapt and work more efficiently with the little time I do have. Sometimes even I wonder how I survived going to morning classes, afternoon lab work, and evening rehearsals for my music major, but if I can do it then so can anyone else.

Melanie, Emory University, Goldwater Scholar

Healthy Lifestyle for a Smarter You

Physical fitness is not only one of the most important keys to a healthy body, it is the basis of dynamic and creative intellectual activity.

John F. Kennedy

Effective time management isn't just making enough time to study; it's also about allotting time for all the other essentials of a healthy life. Your body is like a car: the better you maintain it, the smoother the ride will be to your destination. Any goal will be easier to tackle if you have taken care of your mental and physical well-being.

Studying will demand a significant chunk of your time in college, especially as a STEM major. However, don't forget to maintain a healthy lifestyle each and every day. Now is the time to start getting into healthy habits for the rest of your life, and this section explains why.

Get That Heart Rate Up

Doing well in your science courses requires more than just long study-marathons at the library. While it may seem intuitive that devoting the full extent of your time to your work will improve your academic performance, physical exercise can provide a powerful boost to your studies. A 2011 article published in the *Journal of Applied Physiology* demonstrated that college students who exercised before class improved their test scores by an average of 17%.[3] Moreover, exercise is correlated with higher IQ as well as better concentration when studying. While taking a break from the library to exercise may seem like a chore, the benefits to your schoolwork are real.

The US Department of Agriculture's Choose My Plate program recommends two-and-a-half hours of moderate aerobic exercises or one-and-a-half hours of vigorous aerobic exercises each week spread over at least 3 days.[4] Whenever you feel your attention slipping away, go out for a quick exercise. Compared to taking a nap or surfing the Internet, working out will let you feel more refreshed and more focused on your work once you return to your books.

In university gyms, you can always find students poring over their notes on treadmills or between sets. As we've mentioned before, people aren't as productive with multitasking as they believe, and letting your attention drift while pumping your muscles is a good lead-up to getting hurt. To get the most out of exercise, treat your workout as a break from work. In any case, with good time management skills, you won't even need to multitask. If you really are short on time, however, you can squeeze in some low-intensity studying, like listening to an audio recording of your lecture.

Eat Well

The stereotypical college student is prone to eating oily and salty meals—like mozzarella sticks or instant noodles—at odd hours. However, nutritional research links healthy eating habits with excellence in the classroom. More specifically, one study found that students who consumed the proper portions of vegetables, fruits, protein, and fiber performed better academically than their peers who did not.[5]

Eat three or four well-balanced meals every day to give yourself adequate mental fuel. We know you've heard this before, but breakfast is an especially important meal. In the morning, your body is starved of the nutrients it needs to operate at peak performance. If you don't have the time, grabbing a granola bar or a fruit on your way to class is better than nothing at all. Doing so will help you wake up and have enough energy to last until lunch.

Finally, limit your alcohol intake. The Harvard School of Public Health College Alcohol Study found that students who binge drink—defined as consuming "five or more drinks in a row for men and four or more drinks for women on one or more occasions during a 2-week period"—were more likely to miss class, fall behind on coursework, and get a lower GPA.[6] Some students may consider drinking to be a significant part of their college experience, but getting drunk or hungover on a regular basis won't do any favors for your academics.

Get a Good Night's Sleep

Sleep is an absolute necessity for doing well in college. One student whom we interviewed, Hongyu, a Goldwater Scholar from Dartmouth College, had this to say on the subject of sleep: "I stuck to a strict no-work policy after 10 p.m. This gave me pressure to finish work during the day, which improved efficiency, and led to fun, stress-free nights and healthy amounts of sleep." Getting enough sleep every day is essential for maintaining health and being productive in college. Unlike eating and exercising, poor sleep will affect you *quickly*. Heavy course loads may demand extra effort and in some cases even necessitate an all-nighter. However, try your best to set aside an ample amount of time each evening for sleep. Aim for roughly 7 to 8 hours of sleep a day, the amount the Centers for Disease Control recommends for adults.[7]

Research (and hopefully common sense) suggests that sleeping adequately has significant benefits to both mental fitness and health. A study on undergraduates found that frequency and quality of sleep were among the main predictors of academic success, along with class attendance and past academic achievement.[8] In fact, sleeping only 4 to 5 hours per night or going 24 hours without sleep results in a mental handicap similar to having a blood alcohol content of 0.1%.[9] Consider that in most states, a BAC of 0.08% or more qualifies as legal intoxication. If you are sleep deprived, you won't be able to effectively absorb information, so, in reality, by investing in your sleep, you'll be saving time.

In an effort to stay up late into the night, many of your

peers might develop a dependency on soda, coffee, tea, or one of the many super-caffeinated energy drinks marketed as study aids for stressed-out students. If you can, get some sleep instead. When your body tells you that you're tired, you should listen to it. Taking a quick nap or going to sleep a little earlier than usual is a much more sustainable and effective way to stay awake and focused later on.

Make Time for Fun

In addition to leading a healthy lifestyle, find time for fun. As important as being disciplined with your schoolwork is, it's equally important that you make time for social and extracurricular activities.

You will be happier if you step away from your coursework now and then. Happier students tend to be not only more successful with their work but also more excited and enthusiastic to learn—they enjoy being students more. Also consider that your academic success will rarely ever be considered in isolation. In other words, no one—employers, graduate programs, your future boss, etc.—wants an academic automaton with a 4.0 GPA. They want a balanced, intelligent, and motivated individual who enjoys the things she does.

Cultivate connections and skills outside of your coursework by exploring the student groups and activities offered by your school—from club sports and volunteering to music ensembles and campus publications. College is a time to develop your existing interests and try out new ones. There are no "right" activities, only ones you will enjoy and find fulfilling.

But don't go overboard with clubs, especially during your freshman year. As we've mentioned, you need to be stingy with your time. Many college organizations operate with little to no oversight, meaning that students take on the bulk of the responsibility for arranging meetings and events. In addition, clubs will work at a much higher level than what you may be used to. For instance, while the debate team in high school may have been composed of enthusiastic amateurs, the debate team in college may be largely composed of distinguished veterans from their respective high school teams, many of whom are interested in careers like politics that require public speaking skills. Joining a student group in college can be a bigger responsibility than you may realize, so don't overcommit yourself, and say no if you need to. Participating in a couple of extracurricular activities that you are most passionate about is the best way to go.

Take Care of Your Mental Health

Even changes we want for ourselves in life bring fresh challenges, which require new skills and adaptations. Living on your own, working hard, adjusting to a new social environment, competing with peers, and planning for your career can be stressful even when you are happy and eager to be heading off to college. In short, college is stressful for everyone and even more so when some of the changes are unwelcome or even daunting.

The purpose of this book is to help you succeed as a science student. Anything that interferes with your ability to make the most of your time in college needs your serious attention. Your mental well-being is a frequently overlooked factor in your success. Most of us feel surprised to learn that 80 percent of college students surveyed report feeling "overwhelmed" by all they had to do in the past year and 45 percent have even felt things were "hopeless."[10]

Our surprise is symptomatic of a crucial problem for college students: We are often ignorant of the degree and severity of stress that other college students experience because they are all too often endured silently and with a feeling of aloneness and stigma. Keeping stress to yourself makes more stress, which can contribute to a vicious cycle culminating in problems that can interfere with your academic and social life. More than 25 percent of all college students in any given year have been diagnosed or treated by a mental health professional, and a far greater number describe depression and anxiety as among the top impediments to academic performance.[11] Fully half of all college students report feeling that anxiety is interfering with their academic success and 31 percent report feeling so depressed that it is difficult to function.[12]

There are things you can do to minimize or prevent pressure from blocking your success. First and foremost, take good care of yourself. But sometimes that's not enough. If you find that stress in college is causing you suffering or interfering with your academic or social functioning, it is imperative to break your silence and seek help. This could involve speaking with friends and family members or taking advantage of your college's student health services or services in the surrounding community. You are not alone, and the worst thing you can do is to ignore your stresses and try to carry on.

Also, avoid socializing with people who tire you out, worry you excessively, or generally take your energy ("How do you think you did on the quiz? Did you get the same answer to #41 as I did?"). This will keep you more upbeat and focused on your success rather than weighed down by the expectations of others.

Students Say: What Advice Do You Have for Having a Fun and Productive Time?

Be happy. Fill your life with things that make you happy. Efficiency drastically increases with happiness, along with sleep and a full stomach. Time management is important so you can have time to do the things you love without compromising sleep and food.

Xiaotian, Dartmouth College

In my experience, physical activity best combats stress. Even during exam period, after studying all day, I would always go to the gym before it closed at night. In addition to the well-documented positive effects exercise has on one's mental and physical state, working out also gave me a moment of respite from all other pressures. The hour I spent working out made the next four hours twice as productive and half as stressful.

Chris, Amherst College, Goldwater Scholar, Churchill Scholar

Get involved with extracurricular activities that you enjoy; don't lose track of your passions. Surround yourself with people who can help you out and make you feel great.

Vanessa, University of California, Riverside

Take time to be with friends—it is great to have a support network and to have fun! But also keep in mind you are in college to get an education and are paying a lot for that, so keep your partying in check.

Sara, Yale University

Lots of students like to brag about how little sleep they get, as if it somehow legitimizes their effort. Don't join in their game. Know how much sleep you need, and get enough every night. If you prioritize your sleep, your grades will show it.

Max, University of Minnesota, Goldwater Scholar, Churchill Scholar

Conclusion

College is more than a degree factory. For many, it's a place to find out more about themselves, explore their interests, and learn about the wider world. You will be inundated with opportunities, and you will probably find yourself juggling obligations to your family, friends, academics, and other activities. College will be exciting and rewarding but also mentally and physically draining at times. Exercising, eating right, sleeping well, having fun, and tending to your well-being will help you get the most out of your student experience.

3 How to Excel in Your STEM Courses

I have never let my schooling interfere with my education.

Mark Twain

In chapter 2, we covered how you can spend time to get the most out of college, and now we'll explore how to study like a pro by telling you a number of tried-and-true methods used by successful STEM students to get the most out of classes, laboratory sections, and study time.

While you will have spent more time in your high school classes than you will ever spend in university classrooms, college lectures pack a much greater volume of study material into each hour, and assignments will be harder and take longer to complete. Overall, college courses will demand a greater effort, and what was adequate or even very useful in high school may not be enough.

Think of your college professor as the conductor of an orchestra—she directs your learning while leaving the specifics of performance and practice up to you; it's your responsibility to figure out how you learn most effectively both in and out of the classroom. Fortunately, we're here to help! If you follow the advice in the upcoming pages, you will be well prepared to learn efficiently and excel in your lectures, labs, examinations, and independent study.

Be Preemptive with Your Courses

Genius is initiative on fire.

Holbrook Jackson

If you want to pass, the bare minimum you need to do is to keep up with the course material. If you want to excel, you need to go above and beyond what everyone else is doing. Roger, a biology major from Wake Forest University, explains how he was able to get the most out of his college experience by taking initiative in his coursework: "Instead of asking, 'What do I need to do now to not fall behind?,' ask yourself, 'What can I do now to get ahead?' There is almost always something to be done. Once you've finished working, you will be relaxed and more able to enjoy your down time." In order to do your best in your courses, take a preemptive attitude regarding your learning style, your workload, and your resources.

Step 1: Understand Your Learning Style

Sun Tzu was a Chinese military general, a philosopher, and the author of *The Art of War*, a treatise on warfare. You've likely heard of this oft-quoted passage in one form or another:

> "If you know the enemy and know yourself, you need not fear the result of a hundred battles. If you know yourself but not the

enemy, for every victory gained, you will also suffer a defeat. If you know neither the enemy nor yourself, you will succumb in every battle."[1]

So what does a military general from the sixth century BC have to do with your STEM courses? Well, Sun Tzu never had to cram for a final, and you'll (probably) never have to plot the destruction of thousands of soldiers on the battlefield, but to conquer any trial, you should understand the nature of the problem and how you can mobilize your own strengths to overcome it.

Everyone has a different learning style, and you probably have some idea about what worked or didn't work for you in high school. If you don't know where to begin, take a formal assessment of your learning style. Your university academic center will offer evaluations like the Learning and Study Strategies Inventory (L.A.S.S.I.) or Felder's Index. Some students find these tools useful to plan out optimal study strategies and assess their skills and weaknesses. Take the initiative to think about how your brain works.

Different courses—even ones in the same major—will require a different study approach: flashcards, lecture outlines, highlighting, practice problems, group work, etc. In a biology course you may think, "Everything I need to know is on the lecture slides, but the slides don't have enough context for me to see the whole picture. So, I'm going to skim through the textbook, read over the concepts that I don't understand very well, and then focus on committing each lecture slide to memory." In a physics course, you may elect to take a different approach, "I'm going to review my notes, retrace the steps that the professor took to arrive at the final formulas, and then devote most of my time to solving practice problems."

This is what we mean by studying deliberately. Instead of trying to cram information into your brain without a game plan, figure out what information matters the most and how you can make it your own. Experiment with new study methods offered throughout this chapter and find what works for you.

Step 2: Know What You Need to Know

Again to take a leaf out of Sun Tzu's book, in order to overcome any academic challenge, you need to understand what the challenge entails. At the beginning of every course, develop an idea of the course expectations by reading over the syllabus, which lists the due dates of assignments and tests, as well as the breakdown of how you will be graded. Understand how each test, quiz, assignment, and project contributes to your final score. Usually, the professor will provide a percentage-based breakdown of how your academic performance will be evaluated (e.g., midterms 1 & 2 = 60%, homework = 10%, final = 30%).

Incoming college students often underestimate the role of professors in the quality of a given course. Each professor has his or her own teaching style. Some will be difficult graders of problem sets; others will not. Some professors may elect to put out exceedingly difficult exams with forgiving curves; others may draw problems directly from the textbook. Talk to students who have taken a class from the same professor in the past. By understanding what will be expected of you, you will be able to focus on studying the high-yield information— what matters most in your course.

Step 3: Identify Tools and Resources That Can Help You Succeed

Often, students wait too long to gauge how they are doing in class—they wait for the first exam, the first lab report, or the first paper. Don't wait to be jolted by your first bad grade to get help. Instead, figure out the resources available to you at the beginning of each term. Don't expect that someone will see you drowning and throw you a lifeline; *ask* for help from professors, teaching assistants, advisors, and classmates. Attend your professor's office hours to determine how best to improve your performance. You will not be the first student that they've seen struggle, and you will not be the last. In fact, your professor may be able to direct you to some helpful strategies. Have a chat with upperclassmen who have already taken the class to see what their experience was like. More often than not, they will be able to tell you something useful about the professor's style, the class expectations, the most important concepts to study, or even a textbook or website you have never heard of that will be useful for the course. Your university may also have other resources available to students, such as peer tutoring and lecture transcriptions.

Tackle the Lecture with a Plan

Eighty percent of success is showing up.

Woody Allen

Step 1: Before the Lecture

The first rule of lectures is to go to lectures. The second rule of lectures is to go to lectures. The lecture will be the first and best introduction to the material you need to learn. Unless your classes are recorded, you will only have one chance to see and hear them presented. Rain or shine, sleet or snow, show up for *every single class.*

The act of getting up, brushing your teeth, walking to class, and sitting down in the lecture hall will also make you more motivated to pay attention to the lecture than you might be if you were on your bed watching the video of it on your laptop. Attempting to learn material for a lecture you missed requires more time and effort than going to class in the first place. Additionally, studying on your own becomes much easier if you have already exposed yourself to the material. Just showing up already puts you ahead of those who are sleeping soundly in their dorm rooms.

If you need any more reason to attend your lectures, consider the money that you are paying for your college education. At the time of writing, the average cost of tuition and fees for the whole school year was over $30,000 for private schools and $9,000 for public schools—prices that have since undoubtedly increased.[2] After a little back-of-the-envelope calculation, we found that if you skip out on a two-hour class, you've shelled out about $125 (private school) or $37.50 (public school) worth of college instruction for those extra Zs. This doesn't even take into account the value of the time that you'll have to spend on your own to keep up with the work that you've missed.

Do the Math!

(Tuition per term) / (Number of hours of class per term) = Cost of tuition per hour

(Cost of tuition per hour) × (hours per lecture) = Cost per lecture

Write this number down and post it on a visible location and stare at it whenever you feel compelled to skip out on a college lecture.

When you find yourself at the lecture hall, sit in the front row. College introductory classes can be huge and impersonal. If you sit up front, you will feel more engaged and the fact that you are looking directly at the professor will make you less likely to doze off. You won't be distracted by students sitting in front of you as they play games, message their friends, and check social media. Prevent other distractions by not sitting next to individuals who are not paying attention to the lecture, want to socialize, or seem disinterested in the subject material. From the moment that your professor begins to speak, your only concern should be what the professor is trying to convey to you.

Come to class prepared. Most professors will recommend that you preview the material before coming to lecture. The preview isn't as essential as attending the lecture itself, but skimming through the preparatory material may help you get a general sense of what the lecture may cover (more on this later in this chapter).

Step 2: During the Lecture

Note taking makes you listen actively to the lecture and provides essential study material for your tests and quizzes. When you take notes, your goal is to record general concepts and the relevant details that support them. Summarize the points; don't copy everything that your professor says verbatim. Not only would this make your notes too long and difficult to decipher while studying on your own, it would keep you from processing the lecture itself.

Focus on how each concept fits into the lecture. Why is this important? How does this information fit into the context of the concept that came before it and the one that followed? How could this information be tested? Sara, a molecular biophysics and biochemistry major from Yale, recommends that students constantly try to figure out how each nugget of information fits into the bigger picture: "Always ask yourself why the information being presented is relevant or important to the overall lecture as a whole. This will encourage you to pay attention in class, making it easier to learn the details later on."

Jot down questions and concepts that you think you should review later more thoroughly. If you want to ask a quick question during class or you've noticed that the professor has made an error, don't be afraid to ask; someone else is wondering the same thing. If you have more questions, ask the professor after class while it's still fresh in your mind, or revisit the concepts later on your own.

Taking Notes

Taking notes on a laptop is great for biology, neuroscience, and other content-heavy courses without a ton of special notations. Word, Google Drive documents, and digital notebooks (see "Develop a Thorough Understanding of the Material" below) make it easy to edit, save, and share. Some students take audio recordings of their classes and use these later to fill in any gaps they may have in their notes.

Use paper or a tablet to take notes in courses that involve diagrams, equations, and practice problems like physics, mathematics, and chemistry. Consider using a four-colored retractable ballpoint pen (black, blue, red, and green). The multiple colors are a great tool to make your notes clear and digestible.

For instance, in chemistry class you can more easily transcribe chemical reactions by using red for electrons, blue for arrows, and black for chemical structures.

Note-Taking Techniques

You've been introduced to various note-taking methods throughout middle school and high school. Most students use a variation of the Cornell Method, pioneered by Walter Pauk in the mid-twentieth century.[3] In this format, students divide their papers into two or three columns. Students write their lecture notes in the second column and reminders to themselves in the optional third column. Immediately after the lecture, students review their notes and fill out keywords—also known as "cues"—in the first column to organize their lecture notes into different chunks. This helps students to reprocess the information and organize the notes for later. It looks something like Table 3.1.

Another common note-taking technique is the indented outline. Write the main points nearest to the left margin. Then below, indent and write out the supporting details, such as definitions and examples. You can also expand on the supporting details by indenting again, and so on and so forth.

September 5, 20XX—Biochem—Lecture 2

I. Molecular complementarity
    a. Components can fit together and function together.
    b. Governs how proteins interact

II. Linking/breaking of amino acids in polypeptides
    a. Link: condensation reactions
    b. Break: hydrolysis reactions

III. Secondary structure

    a. The local folding of a polypeptide chain into regular structures

        i. Includes: alpha helix, beta sheet

IV. Alpha helix

    a. Common secondary structure

    b. Linear sequence folded into a <u>right-handed spiral</u> stabilized by hydrogen bonds between <u>carboxyl and amide groups</u> in the backbone.

    c. Important structure for integral membrane proteins, s.a. channels

    d. Stabilized by non-covalent bonding.

Table 3.1. September 5, 20XX, Biochem, Lecture 2

Keywords/Cues	Notes from Class	Note to Self
Molecular Complementarity	- Molecular Complementarity: Components can fit together and function together. - Governs how proteins interact	Confusing, <u>ASK THE PROF.</u>
Linking/Breaking of amino acids in polypeptides	- Link: <u>Condensation</u> reactions - Broken: <u>Hydrolysis</u> reactions	Draw out the two reactions later. <u>Probably will be on test.</u>
Secondary Structure	- Secondary Structure: The local folding of a polypeptide chain into regular structures including the alpha helix and beta sheet.	
Alpha Helix	- Alpha Helix: Common secondary structure - Linear sequence folded into a <u>right-handed spiral</u> stabilized by hydrogen bonds between <u>carboxyl and amide groups</u> in the backbone. - Important structures for integral membrane proteins, s.a. channels - Stabilized by non-covalent bonding.	Isn't there a figure in the textbook for this?

Finally, if your professor provides slides before lectures, consider writing your notes directly on the PowerPoint file or on printouts. This will make it easy to refer to the slides while studying and is particularly useful for diagram-heavy courses.

However you take your notes, keep them organized so that you can pull them up later. Record the date and the topic of the lecture on top of the paper or in the name of the file. Then store your notes in an easily accessible location, such as a digital file, an expanding folder, or a three-ringed binder so that you won't have to fish out sad, crumpled sheets from the bottom of your bag.

Step 3: After the Lecture

As soon as you can after class, review your notes. Organize your notes to make them coherent and, more importantly, mark the concepts that you didn't understand. Look through your notes while the lecture is fresh in your head to fill in any gaps in your understanding by reading the textbook, looking up the information online, asking your peers, or going to office hours.

Say you are given two sequences of letters. One sequence is completely random (e.g., uqkbregyxu). The other sequence forms a word (e.g., activation). Which one is easier to learn and remember? The second, of course. This is because there is an underlying logic that you can use to recall the information contained in this sequence. In science, no information is truly random or isolated. The facts you learn always form a "word" in some sense because they fit into the larger web of concepts that scientists use to organize the natural world. For instance, the hundreds of reactions you will learn in organic chemistry are all based on key principles that make them understandable. The seemingly random membrane channels, organelles, and proteins in a cell seem overwhelmingly complicated when learned by each component, but make greater sense if seen as a cohesive whole.

Appreciating how scientific facts are related to one another demonstrates true scientific understanding, not just knowledge of the facts themselves. A rookie mistake that many new students make is only memorizing the information they are given. This is a poor way of learning science; not only does it not accurately reflect the essence of science, but also it's an incredibly inefficient way to pick up new facts. As you begin to study, focus on figuring out the general theme of how the lecture is structured and then fill in the holes with relevant details.

To sum it up, this is how to approach each lecture:

1. Before the lecture
    a. Choose note-taking tools and techniques appropriate for your course.
    b. Briefly review the material that will be covered.
2. During the lecture
    a. Think about how all the information can fit into a cohesive whole.
    b. While taking notes, summarize. Don't transcribe.
    c. Ask questions, either during or after class.
3. After the lecture
    a. Organize and mark your notes.
    b. Review the concepts that weren't immediately clear to you during the lecture.

Read Your Textbook Deliberately

When eating an elephant take one bite at a time.

Creighton Abrams

If your professor assigned any reading for your class, go over it, but treat your textbook as a secondary resource—something to supplement your lectures, not replace them.

So, how should you approach the reading? The answer is—always—deliberately. Like attending lectures, don't ever treat reading like a passive exercise. We've all had moments when we felt our eyes glaze over the pages of our textbook, letting the words appear on the retina without registering them in our brains. By the end of your reading, you may have flipped through the chapter, but did you absorb anything from it? Probably not, and this is precisely what you want to avoid.

Step 1: Pre-read

Before you focus on your reading, quickly read through it once. Note any headings, subtitles, diagrams, examples, figures, and equations and whether any of it could help you flesh out the questions that you have from your lectures. A quick read-through gives you a glimpse of what the text will cover, where it will take you, and how important it is to the lecture. Having a taste of the reading will help you to digest it more readily on your second, deeper read-through.

Step 2: Read

Now that you have some idea of what the reading will cover, it's time to fill in the gaps. If you are in a detail-heavy STEM course, use the text to reinforce and add to your lecture notes.

Some information will be more important than other information. Focus on the concepts that were covered in class. If a concept in the textbook wasn't covered in your class, then it probably wasn't important for your professor and will not be high-yield for your test. Underline concepts that you'd like to return to, but avoid excessive highlighting or underlining. If everything is important, then nothing is.

Step 3: Summarize

By the end of each section, summarize three to five key points from your reading and jot them on the margins of your textbook, in your notes, or on a separate piece of paper. This exercise will keep you honest about whether you've actually processed the material. Don't copy down the text word for word; this is another path to passivity. Instead, rephrase the reading into your own words and keep it short and sweet.

Memorization Techniques

To know the laws is not to memorize their letter but to grasp their full force and meaning.

Marcus Tullius Cicero

In high school, STEM courses mostly consist of memorizing facts—picking up the basic building blocks of science. College STEM courses, however, will expect you to pick up those blocks and build a tall tower. Memorization, therefore, is necessary, but not sufficient, to ace your STEM courses.

There is an abundance of fancy memorization techniques out there, but for the purposes of your courses, the simpler the better. Here, we present two memory methods that you are already well familiar with, but with a slight twist:

Technique 1: Spaced Repetition

Spaced repetition is a learning technique that relies on the simple fact that multiple exposures to the same information over time is beneficial for memorizing. This—even without the decades' worth of research supporting it—is common knowledge, but implementing spaced repetition into your study regimen used to take diligence, initiative, and lots of flashcards.

Fortunately, there's an app for that.

You can download spaced repetition apps that organize your flashcards for you based on how well you know them so that the concepts that you are less familiar with appear more frequently than the concepts that you already know. You can make your own virtual deck or download one of the many pre-made decks available online.

Max, a Goldwater Scholar, Churchill Scholar, and neuroscience major from the University of Minnesota, told us how spaced repetition helped him with his courses: "For memorization, use flashcard software like Anki that will bug you when you get questions wrong. I used to struggle with memorizing until I discovered this trick. Taking the time to consolidate the material into flashcards and study them daily, even if only for ten to fifteen minutes a day, will make those facts a breeze on the exam."

Technique 2: Interesting Mnemonics

Mnemonics are memory devices that use associations to help you retain information. You have probably used ones like "Please Excuse My Dear Aunt Sally" to remember the order of operations (Parentheses, Exponents, Multiplication, Division, Addition, and Subtraction) or "SOHCAHTOA" to remember how to calculate the Sine (Opposite/Hypotenuse), Cosine (Adjacent/Hypotenuse), and the Tangent (Opposite/Adjacent) of a triangle. You probably still remember these memory devices—or variations of them—which should demonstrate the lasting power of these verbal associations.

Make up mnemonics to memorize and digest large amounts of information. For instance, take this list of the cranial nerves (CNs):

- O: Olfactory nerve (CN I)
- O: Optic nerve (CN II)
- O: Oculomotor nerve (CN III)
- T: Trochlear nerve (CN IV)
- T: Trigeminal nerve (CN V)
- A: Abducens nerve (CN VI)
- F: Facial nerve (CN VII)
- V: Vestibulocochlear nerve (CN VIII)
- G: Glossopharyngeal nerve (CN IX)
- V: Vagus nerve (CN X)
- S/A: Spinal Accessory nerve (CN XI)
- H: Hypoglossal nerve (CN XII)

Complicated, right? Not only do you have to remember the names for each cranial nerve, but you also have to keep them in order. Fortunately, this is where mnemonics come in. The first letter of the name of each nerve can be used to create sentences like:

- Ooh, Ooh, Ooh, To Touch And Feel Very Good Velvet. Such Heaven!
- Oprah Ought Order Tasty Treats And Finally Value Growing Voluptuous And Happy

You can decide which mnemonic is easier to recall, but generally, students are better at remembering a voluptuous Oprah than less interesting images. The more strange or obscene it is, the better mnemonic sticks in your mind. This phenomenon is known as the Von Restorff effect; you are more likely to remember something that "stands out like a sore thumb" than something more mundane.[4] To create especially memorable mnemonics, incorporate you, your family, friends, and pets in situations that you'd never, ever expect in real life. The more colorful, bizarre, embarrassing, and cringe-worthy, the better you will recall the mnemonic.

Furthermore, you are the beneficiary of decades—if not centuries—of frustrated students, and you will be able to find many clever mnemonics archived online for you to adapt into your own study regimen.

Just remember that memorization is only the tip of the iceberg. There's no point in memorizing the order of the cranial nerves if you don't understand their role in the body. By itself, memorizing will never make up for actual understanding.

Tackling Homework and Problem Sets

Opportunity is missed by most people because it is dressed in overalls and looks like work.

Thomas A. Edison

Step 1: Routinize Your Study Hours

One of the most difficult aspects of tackling your homework is actually sitting down and making time for it in your busy college schedule. Sam, a Rhodes Scholar and Goldwater Scholar from the University of Chicago whom we interviewed, suggested to us that students make studying a habit by doing homework and studying for courses at the same time every week. "This way, I made sure that things didn't get too rushed right before assignments were due and that I had time to go to office hours if I needed help with any problems. I also didn't cram for exams and instead tried to study a little bit each day over several days. This allowed me time to absorb and understand the material."

Step 2: Treat Your Homework as a Tool

Practice questions and problem sets are staples of STEM courses. Think of these problems as tools to gauge your academic mastery, like the strength tester that you might find in fairs. The bigger your conceptual muscles, the better you can strike at the heart of the problem. Having trouble with questions should indicate your weak areas, which you can then focus on improving.

The point of solving questions is to figure out what you need to review more of, so don't burn through your questions all at once. Gain some understanding of the material being tested before trying to solve them. Treat the questions as a mini-test, and do the problems without peeking at the answers. If you can't answer a question, move on. When you finish, consult the answer key and figure out the underlying logic behind each question. Ask yourself:

- What concept was the question trying to test?
- How did the answer key solve the question?
- Do the steps taken to solve it make sense to me?
- Does the answer make sense to me?

Step 3: Review What You've Missed

Mark every question that you missed or couldn't answer. Figure out why you made each mistake, and compile a master checklist of the concepts and ideas that you need to review to perfect your understanding. Just glancing over the answer key is not enough; fill in the holes in your knowledge by re-reading your lecture notes and textbook, asking your classmates, or talking with your professor. Once you gain a better understanding of the concept needed to solve the problem, return to the questions that you've missed and try again to answer them.

Continue this cycle until you are comfortable with answering your problems. *All of them.* There's a reason that your professor assigned these questions, and you can bet your tuition money that many of the ideas covered in your homework will appear on the actual exam. It's much better to struggle with a problem in the comfort of your room or the library than to meet it for the first time in the exam room.

Develop a Thorough Understanding of the Material

If you don't know where you are going, you might wind up someplace else.
Yogi Berra

You may be planning well and putting in lots of study hours, but how do you know that you are actually getting the most out of your classes? The following are some tips to ensure that you really understand the material you've studied and are maximizing the utility of your class work.

Tip 1: Would You Be Able to Explain It to Someone Else?
Mark, a physics and Earth sciences major at Dartmouth College whom we interviewed, made this point to us about how students should check to see if they are learning their course material: "If you can't explain in fairly simple words the concepts you're trying to study, you don't fully understand them."

If you can't explain what you are learning, you don't really know it. To consolidate your information, draw out pictures or diagrams, or say it out loud. What is the polymerase chain reaction? Someone who has a poor understanding of the concept might be able to regurgitate some details, like saying that it is a chemical reaction that employs strands of DNA, nucleotides, primers, and enzymes. Someone who does understand it will be able to explain to any intelligent nonscientist that the polymerase chain reaction is a technique to copy a specific sequence of DNA. You should be able to distill the complicated ideas you learn in the classroom into simple explanations.

Tip 2: Ask Questions, Especially to Yourself
Always, always, always ask questions—in class, to yourself, and to your peers. According to Paloma, a biology major from Williams College, "Don't be shy, ask questions. Think deeply about what you are learning and engage with the material so that it truly makes sense to *you*. Be fearless in your search for answers. Once you are in college, it is less about grades than it is about truly understanding concepts." When you ask questions, you are challenging your current understanding of an idea so that you can improve it. Even if you feel pretty comfortable with your understanding of the material, continue

to ask yourself, "What would happen if this step went wrong?" or "Why is this idea or step or process important?" or "How could this be applied elsewhere?"

Tip 3: Attend Office Hours and Review Sessions

Your instructors have spent years, even decades, to arrive at their current depth of understanding of their discipline, and each week, they will devote a few hours to answer questions about their courses in office hours—a fantastic academic resource often underutilized by students.

Office hours are also a great way to get to know your professors and/or graduate student teaching assistants. Demonstrate initiative by showing how you've tried solving the problem and by identifying the steps that confused you. Building a relation with your professors is a great way to receive good letters of recommendation for jobs, scholarships, or graduate school applications. A faculty member who has never gotten to know you beyond the fact that you took his course can do little more than affirm that you took a course and got a good grade in it. This sort of recommendation has little value since your grade is already listed on your transcript. By going to office hours and asking questions, you have the opportunity to demonstrate your drive and your enthusiasm.

Introductory and intermediate-level courses will typically offer review sessions before a midterm or a final. This is the chance to clarify any last-minute questions with the teaching assistants or the professor, and going over the toughest concepts again might help you see them from a new angle. Plus, watch out for hints and clues about what might be especially high-yield on the test.

Tip 4: Work with Others

Some students benefit from studying in a group. Your group study session is only as effective as the members, including you. No one wants to work with someone who doesn't pull his weight, so pick study buddies who take their work seriously and can inspire you to work even harder.

A group study session is an opportunity to refine your understanding of the course material by bouncing ideas off your peers and checking to make sure that you really grasp the concepts from class. You can help keep distractions at bay by preparing an outline of what the group should go over during a study session. If you run into a disagreement, ask your study buddies to explain their reasoning, step by step. Either you will help them spot a fault in their logic, or you will learn something you hadn't thought of before.

Aaron, a biology major from Dartmouth College whom we interviewed, shared these thoughts with us: "Particularly for classes with ungraded problem sets, keeping up with material is challenging. Not only does group study make me accountable to my classmates for consistent preparation, but it also helps my time management by scheduling sessions without the distraction of work from other classes."

Tip 5: Take Advantage of Apps and Online Resources

There are very few subjects that you cannot find and learn online. In 2002, the Massachusetts Institutes of Technology (MIT) announced that it would post materials from many of its courses on the Internet for free, leading to the creation of MIT OpenCourseWare. This was the beginning of a movement among many institutions to start Massive Open Online

Courses (aka MOOCs) and bring free education materials online through companies like Coursera and edX. One of the very best places for free online learning is the nonprofit website Khan Academy, which uses short videos and interactive exercises to teach a wide variety of STEM and non-STEM subjects to viewers.

Online courses and Khan Academy are just a few examples of online learning materials that you can use to complement your education. These classes will not give you college credits, but they can help you to review and to expand your knowledge. You may find yourself in a class or a particular lecture that is taught poorly, and these resources might make the difference between learning and struggling.

Learning online may be the way of the future, but for the present, you do need to go to class. Here is why: you can't ask your questions to a recording, and if you depend too much on online content, you might find that you are learning different things than the students who went to class (unless you are watching recordings of your own class's lectures). Plus, if you have questions when you are watching a video online, you don't have much recourse for getting them answered immediately. Lastly, when it comes time to ask for a letter of recommendation from a professor or when you're in the working world and need a network of scientific colleagues to help you out, you will be glad you took the time to meet your professors and classmates in person.

There is a never-ending list of learning tools, software, and apps that you can use to help you with your courses. Here are a few productivity and study tools that you may consider checking out:

- As mentioned before, spaced repetition software like Anki and Memrise generate flashcards and quiz you on the concepts you have the hardest time digesting.
- Wolfram Alpha is a powerful tool for doing computations online.
- Quizlet has gigantic question banks made by students around the world.
- Evernote and OneNote are digital notebooks to create notes, clip content from the web, make your content easily searchable, and sync all of your notes through all of your devices.

Virtual resources change quickly. In fact, they probably changed a little bit as the ink was drying on this page (or as the page was loading on your tablet). What won't change, though, is the need to keep an eye out for what is new, what can make your job as a student easier, and what can be used to supplement your courses.

Tip 6: Stretch Out Your Study Time
Cramming just might barely save your butt in an exam, but reviewing as you go helps you understand the material and retain what you learn for the long term. Cover your course material more than once before you are tested on it. This means that you should plan to review the material for an extended period of time—a method of learning also known as "Distributed Practice." By making time for a second or third review, you will strengthen your understanding and improve your ability to recall specific information.[5]

Tip 7: No Pain, No Gain
Studying and committing new information to memory is like training a muscle; there's a considerable amount of pain and

strain for the gain. You can't go from scrawny to brawny in a single workout session; likewise, you shouldn't expect to understand, memorize, and be able to apply complex concepts after a single cram session. Whether physically or mentally, bulking up will take sustained time and effort, and the sooner you commit these facts to heart, the better you will be able to persevere through the academic obstacles ahead of you. In the words of one of the students whom we interviewed, Paloma from Williams College: "Science is hard. And that is amazing. Fall in love with a subject and learn as much as you can about it. But remember that if you are struggling, you are learning. There is nothing more satisfying than to master something that you previously struggled with."

Students Say: What Academic Tips Have Worked for You?

Find the best professors; their passion will excite you, their thoughts will expand your worldview, and their courses will teach you principles, not just facts.

Chris, Amherst College, Goldwater Scholar, Churchill Scholar

I would go through my notes after every lecture within 24 hours to make sure I understood everything. Anything I could not figure out on my own, I would take the time to meet with the professor to figure it out. When it came time for memorizing things and learning concepts, I used a big whiteboard and drew all the processes until they all made sense. I would start studying for tests about a week before test day.

Roger, Wake Forest University

I think the key to studying is making an effort to enjoy all the material that you encounter. Even in classes that were not the most interesting or heavy in rote memorization, I have always made an effort to place what I am learning in a larger context and tried to link the new information with what I already knew in the field. Not only do those connections make it easier to understand and remember the new concepts, but the big picture view makes the whole process a lot more interesting and exciting.

Peter, Johns Hopkins University, Goldwater Scholar, Rhodes Scholar

Go to office hours—even if you don't have questions going in. If you are engaged during office hours, you can really enhance your understanding by delving deeper into topics that help solidify the foundational material.

Alvin, Harvard University

Don't just attend your classes; be engaged in them. Actively listen and take notes with the aim of understanding the material and making it your own, rather than simply memorizing it. Even if your professors rely on notes more than the textbook, read your textbook anyway; it will provide you with a deeper understanding of the material.

Alyssa, Hope College

Preparing for Exams and Quizzes

Some advice: keep the flame of curiosity and wonderment alive, even when studying for boring exams. That is the well from which we scientists draw our nourishment and energy.

Michio Kaku

Exams will typically be the biggest factor in determining your grade in college science courses. You would probably think that that making a concerted effort to study the information from your lectures would put you in good shape to take your exams. However, this is not always the case, as test preparation and test taking are skills in their own right. With this in mind, you can prepare for high-stakes tests by taking these three steps: 1) familiarize yourself with how you will be tested, 2) mobilize your knowledge during your test to the best of your ability, and 3) use your graded test to improve your performance on future tests.

Step 1: Preparation
Understand What Will Be Covered and How It Will Be Tested

Spend at least a week preparing for a midterm or a final. If you've followed our advice on time management and study skills, you should have kept up with your lectures and should have a good grasp of the course material already. As the exam date approaches, you'll be ready to shift your focus from actually learning the content for the first time to solidifying your knowledge for the test.

Exam-based STEM courses will often provide practice exams or sample problems, adapted from tests administered in previous years. Just like homework questions and problem sets, practice exams are tools to test and refine your knowledge.

You should take practice tests under real exam-like conditions. If you're going to be in a noisy hall when you're required to take the real test, take a bunch of practice tests or solve problem sets in a noisy hall. In addition, take a practice test with the mentality that the stakes for the practice run are real. Recreate the stress levels of your testing environment. Your goal is to be able to recall and analyze the relevant information while under pressure. As you get practice with this, the stress seems much less intimidating and the answers come more easily.

While going over the answer key to your completed practice questions, think of the following: What format of questions does the professor use: short-sentence, multiple-choice, paragraph questions? What degree of detail does the professor expect out of your answer? Does she test for specific detail, the synthesis of different concepts, etc.? By thinking about how you might be tested, you will be able to devote your time to

studying the material that your professor wants you to understand most.

Master the Material

When you study, focus on the material that you have difficulty
understanding. While this may seem obvious, you'd be surprised at how easy it is to push off the concepts that you may
be struggling with. Don't be content with a marginal understanding of a difficult concept. If you're having trouble understanding certain concepts, chances are your classmates are as
well—this places you in the middle of the class curve, if not on
the wrong side of it. Take the time to thoroughly understand
everything that is expected of you, and ask your TAs and professor for clarification.

Make Your Own Questions

Once you really know your course material, you can go the
extra mile and think of questions you believe may be asked
on the exam. In a sense, you are putting yourself in your professor's shoes to anticipate how she will test you. This is the
final step to refine your knowledge.

For instance, Samuel, a Goldwater Scholar and a Rhodes
Scholar whom we interviewed, found it useful to work through
problems backwards: "For example, if one of my homework
problems gave me three numbers and asked me to calculate an
answer from them, I'd give myself the answer and two of the
'givens' and try to work out the third 'given.' This taught me to
think more flexibly when it came time to apply the concepts on
the exams." Breaking down a problem into its disparate parts
helps you to understand what is being tested, how it is being

tested, and how a concept can be applied in different variations of the same problem.

You can also write up your own problems that incorporate challenging concepts or a large number of important facts. This will help you to recall information and anticipate what will appear on your exam in a new context rather than simply skimming through your notes.

Step 2: Test Performance
Confidence, Concentration, Relaxation

Get a good night's sleep before any big exam and eat something nourishing. You need to be running at full mental power for the test, and your brain will be running on a deficit throughout the day if you fail to do either of these things.

Additionally, maintain a positive attitude about the test, and avoid those classmates who want to complain about how they didn't study or how hard they think the test will be. Whether or not you have prepared sufficiently, adding to your stress will only diminish your test performance.

Take the last hour before the test to eat a small snack, hydrate, and relax. When the test begins, bring the entirety of your focus onto the exam. Also consider taking earplugs or a noise-canceling headset to the test.

Track Your Time
Bring a wristwatch or a timer to your testing hall to keep track of your time. Go through as many questions as possible, and don't get caught draining all your remaining minutes on a

particularly challenging problem when you could be earning points on another question. If you are stuck, mark the question and come back to it later. You may find tackling another question will help you return to the difficult question with a fresh perspective and a new plan of attack.

Sit Tight

If you finish the test early, don't leave, and ignore those who do. Always, check and recheck your answers for the remainder of the test period. You are bound to find errors in your answers—a faulty calculation, a bubble filled in the wrong place. These are points saved. Furthermore, trust your gut and think back through difficult or seemingly odd sections of the test. Often, you may find that your understanding of a question was wrong, requiring significant revisions.

Maintain a Positive Attitude

If the test seems more difficult than you had expected, that is not necessarily a reason to be discouraged—maintain a positive attitude and trust your preparation. According to Sara, a Yale graduate, if you hit a roadblock on your exam, "Don't be intimidated and be positive! This is hard—but just focus on yourself and on what you know. Go into an exam thinking about how much you know, not worrying about the things you don't know or find confusing!"

Many professors write exams with questions that are challenging so that they can get a wide distribution of scores. They then assign grades to students depending on where they rank on the curve. Just because you are struggling on a test, does not mean that you won't do well on it. Regardless of how tough it seems, keep working, remain focused, and finish the test.

When you leave the testing hall, it's difficult not to perseverate about how you did or the problems that you've missed, but worrying will do you no good at this point. Do something fun and restorative to reward yourself for your hard work; you've earned it!

Just as some people experience stage fright before a big performance, some individuals experience testing anxiety. A little bit of nerves will probably improve your performance on an exam, but if your anxiety interferes with your ability to do your best, read the section on coping strategies for test anxiety later in this chapter.

Step 3: Postgame Analysis: Figure Out Where You Stumbled

Handing your test to the professor is not the end of your exam experience. Once your test has been graded and returned to you, read through your grader's comments and compare the test to the answer key to see what you've gotten wrong. You should do this for two reasons: First, TAs and professors make mistakes, especially while grading hundreds of exams in a huge introductory STEM course. It's not uncommon to spot a grading error. Second, you can learn from your graded exams to see where things went awry, in the same spirit as an athlete watching postgame replays.

If you missed a question, you did so for a reason. Going through the test allows you to retrace your thought process to see how you brought yourself to the wrong answer. Reviewing your test will lend insight into the *types* of mistakes you tend to make on exams—which is important for improving your test-taking skills. In your posttest analysis, it's not enough to understand the correct answer. Over time, this process of exam review helps you become a more confident test taker. Below is a list of some of the most common mistakes that students make on exams, and how you can begin to fix them.

You Didn't Understand the Concepts Being Tested

This is one of the most common reasons that students get things wrong. They didn't have the time or put in the necessary effort to understand what was being tested. While studying, always ask yourself whether or not you really understand what you are learning and can explain it to someone who has never heard about it before. This will train you to organize the facts that you learn into a coherent whole.

You Misunderstood the Question

Are you reading through the exam too quickly? Maybe you need to take a little bit more time to digest the instructions. If, on the other hand, you were unsure about what the question was asking, ask the professor to clarify during the exam. Depending on the nature of the question, he may or may not be able to help you, but it is worth asking about in case he can.

You Made a Calculation Error

Writing out each step of your calculations on the exam will minimize the likelihood of making a mathematical mistake and, even if you do make a mistake, showing your work demonstrates understanding of the problem that might get you partial credit. Always double-check your calculations, and for each calculation-based answer ask yourself: "Is that number a plausible value?" and "Are the units correct?" If the question is "How many piano tuners are in the city of Chicago?" and your answer is 10 million kilocalories, then this should alert you that you made a mistake somewhere earlier in your calculation.

You Didn't Have Time to Finish

This can be one of the most difficult problems to cope with when taking a test. If you needed more time, this suggests that you still don't understand the material as well as you should. You may need to start working more intensively on practice problems so that you can solve problems more quickly. If almost everyone in your class does not finish an exam, that might be a problem more at the level of your instructor or test writer, but be sure to identify weaknesses in the cases when you're the only one still trying to finish the test after time has been called.

You Had No Idea How to Answer the Question

If there's ever a moment during an exam where you find that you have no idea how to answer a question, *do not* leave the question blank. Unlike the SAT, where your score drops with each incorrect answer, you will not lose points for incorrect answers in college, unless your professor explicitly says otherwise. If you come across a question that completely stumps you, start jotting down concepts that you think may be relevant. Starting to think about the subjects related to the problem may jog your memory and give you insight on how you can approach the question. At the very least, you might get some partial credit for including applicable information, while a blank page will certainly mean no points.

Students Say: How Should Students Prepare for Tests?

Stay focused on high-yield information. Learn what your professor wants you to know.

Ryan, University of California, Los Angeles

Question everything. Especially in proofs, steps that appear obviously true are not necessarily so. Work with other students, even if just to bounce ideas off each other. Try not to wait until the last minute.

Sara, University of Michigan

Do problems over and over again until every single step is crystal clear. You can also talk through concepts with other people to memorize pathways and to understand how experiments work.

Yingchao, University of California, Berkeley

It's all about recall and synthesis—thumbing through one's notes, it is easy to have the illusion that you know the material and can explain it to a peer, or on an exam. But, this superficial recognition is not the same as deep comprehension, or the ability to teach (one of the highest forms of understanding).

Chris, Lafayette College, Goldwater Scholar, Fulbright Scholar

Do a little bit of work on each assignment every night. I found it a lot more relaxing and interesting to draw out the work and mix up different concepts. It is always great to have the luxury of leaving a difficult problem for the next night and letting the understanding sink in, and sometimes, unexpected insights come out of thinking about multiple fields at once. It is also crucial to leave time for extracurricular activities, exercise, and social life, and a more drawn-out schedule means that there is almost never a night that must be fully committed to studying.

Peter, Johns Hopkins University, Rhodes Scholar, Goldwater Scholar

Excelling in Your Laboratory Assignments

Lab assignments let you apply what you've learned in your courses to situations that simulate what you may encounter in research. In general, lab assignments can typically be divvied up in two ways: experiment-based and project-based. Experiment-based assignments will push you to confirm observations that are already well understood in the field. For example, you could do a kinematics experiment using toy cars with sensors in your general physics lab to confirm known relationships between position, velocity, and acceleration or dissect a cockroach to understand the basic physiology of digestive systems. In contrast, project-based lab assignments will ask you to synthesize the concepts that you've learned in class into a new project or knowledge, like taking rock samples of your college's backyard or engineering a user-friendly interface to go through online databases.

Different STEM classes will have their own types of lab assignments, and the physical space of the "laboratory" may range from well-ventilated facilities for general chemistry experiments to your campus quad for observing the spectra of the heavens.

One key difference between high school and college science courses is the fact that lectures and laboratory sections are distinct. In high school, whenever a lab experiment took place, everyone would get up during class, pair off, and go to the back of the room for twenty to forty minutes. In college, however, many science students will have an additional 2 to 6 hours per week *outside of class* devoted to lab sections. Some universities will even allow you to take lecture and lab sections in different terms (e.g., organic chemistry lecture in the fall and organic chemistry lab in the spring). Graduate students and experienced upperclassmen TAs will be supervising the laboratory sections, but your professor may pop in to say hello. In the following section, we describe helpful techniques to prepare for laboratory experiments, how to go through them efficiently, and how to organize your results into a cohesive, successful lab report.

Understand Lab Safety

Before you even walk into your lab facility for the first time, read the relevant documents on laboratory safety and conduct. This advice, of course, is more relevant to labs involving chemicals and heavy machinery. When needed, wear appropriate personal protective equipment such as gloves, lab coats, goggles, or anything else called for given the nature of the experiment. Approach lab safety with an appropriate level of caution.

Be Honest with Your Work and Data

In all experiments, there is a chance that you will not obtain the results you expect. This is a part of the scientific process. However, while real-life research experiments will be repeated and modified as needed, your lab assignments are designed to be finished within the allotted timeframe. Especially in experiment-based assignments, you may not have the time to redo an experiment, but it is your duty as a scientist to report your findings and to address your errors and setbacks. No matter how your findings deviate from the expected results, don't fudge or massage your data, don't copy your friend's work or data, and don't let your friends copy your work.

Fabricating data is the most serious misconduct within the scientific community. Scientists who have been found guilty of doing so in order to corroborate their hypotheses have had their grants, accolades, and jobs stripped away from them. Fabrication goes against the entire point of science itself: building our knowledge of reality from testable experiments. So even if you have strange data, just be truthful with your results. You and everyone else will end up messing up an experiment from time to time. The more practice you have, the better you will get.

Prepare for Your Lab Assignments

How should you get ready for your lab? First off, obtain a copy of the lab manual, read through the directions, take note of anything you don't understand, and ask your classmates and TAs.

To familiarize yourself with the concepts in your lab assignment, look through your class notes and check the references cited by your lab manual. These may be pages from textbooks or science papers that discuss the discoveries that motivated your laboratory assignment. It's always a good idea to find primary sources, both to reinforce your understanding of the material and to later use in your lab report.

Next, summarize the important steps of the experimental procedure clearly and concisely. Rewrite the paragraphs of instructions in the lab manual into a numbered list of steps. In labs involving chemical compounds and reagents, lab instructors may request that you look up any chemicals you use in the lab in a Material Safety Data Sheet (a type of summary of the properties and hazards of a substance) and take note of safety precautions in your lab notebook. Whether or not this is officially required, it is always a good idea to be familiar with what you'll work with during an experiment (e.g., don't touch solid $NaOH$ pellets with your bare hands). Furthermore, you'll be recording a lot of data during the lab, so prepare a table in your lab notebook that you can fill in during the experiment. This way you won't have to do this during the lab when you have a number of other things on your mind.

While preparing for your lab, think about how you can make the most out of your limited lab time. For example, if you have to centrifuge a solution for thirty minutes, what else could you be doing during that period? Could you clean up dirty glassware? Could you weigh out substances that you'll need later? Could you spend your time taking notes on your experiment? If you have done any cooking in the past, you'll find that this type of work style is already very familiar to you—grating cheese while the pasta is boiling, cutting onions as a steak is being seared on the stove, etc. Using your dead time productively will help you to get things done and get you out of the lab as fast as possible.

Finally, be sure to eat well and get a good night's rest before the lab. This will keep you awake and aware of your surroundings as you conduct the experiment. Even if what you are doing is as mundane as pouring one liquid into another, for optimum safety you should be able to put your full attention on the experiment.

To summarize, before each lab, you should

1) Read the lab manual
2) Skim through suggested background reading (journal articles, class notes, etc.) to understand why you are doing the experiment
3) Write a clear, concise bullet point summary of the procedures in your lab notebook
4) Figure out how to use "dead time" productively
5) Understand safety precautions

During the Experiment

After doing all the readings and taking copious notes, you're now ready to ace that experiment. But, uh oh, you've never been in a real lab before! What's a fume hood? Where is all the glassware? Where is everything!? I know I'm supposed to get a P200 micropipette with yellow tips to measure 100 μL of a buffer solution, but I have no idea where anything is!

Fear not, future lab experts! You will find your way. Most students enter college with very limited, if not zero, lab experience. Some of your classmates will seem orders of magnitude more prepared than you, but most will be just as confused as you are. Don't be intimidated.

When starting the experiment, follow the procedures carefully. Start off by putting on any necessary safety equipment, acquainting yourself with where things are, and cleaning your workstation. Make use of the dead time that you have between different steps. If you are working with dangerous materials, be safe and use common sense, and get help whenever you need it. Always ask if you are not sure what you need to do!

Record your observations and all the data you are collecting throughout the lab and write it next to the procedure in your lab notebook. The more detailed you are during lab, the more material you'll have to work with when you write up your lab report. Always write more rather than less. Then, when it is time to leave, double-check that you have accomplished all of your objectives, and clean up your station.

After the Experiment: The Lab Report

Many lab courses will require reports—a summary of the theory, procedure, and results of the experiment. The primary goal of the lab report is to teach you how to record scientific data and results and introduce you to scientific writing. Lab reports may be difficult to complete the first few times, but you will find that they are pretty formulaic and involve rewriting material you should already know—introduction, methods, results, conclusion, etc.

The best time to write a lab report is right after the lab. Since the material is still fresh in your mind, you'll be able to crank out the report quickly. Complete your lab report well ahead of schedule, so that you have time to ask questions or go to your lab TA's office hours when you get an odd result or run into a something that needs clarification.

Cheating, Falsifying, and Plagiarism, Oh My!

Cheating is rampant in college, especially in high-stress majors like STEM. About 75 percent of students have admitted to cheating in college, a figure that has held constant for five decades.[6]

Academic dishonesty, however, devalues your education and that of others. Moreover, it ends up cheating you of the foundation for your future successes. The long-term risks are simply not worth it for the questionable short-term gains. If you are caught, the scarlet letter of academic dishonesty will haunt you beyond your academic career, as graduate and professional school applications will ask about any and all institutional actions you've received.

Troubleshooting for College Students

The difference between the university graduate and the autodidact lies not so much in the extent of knowledge as in the extent of vitality and self-confidence.

Milan Kundera

In the last few pages of this chapter, we discuss strategies for improving your performance if you are encountering difficulties with you courses. The earlier you recognize your weak areas and address them, the better you will do in college. Here are some of the most common roadblocks that college students face as well as tips and techniques to push through them.

I Think the Material Is Boring

Somewhere out there, someone has devoted her entire professional life to the study of a single subject that you've found to be boring. Someone else has devoted years writing a textbook, and a publishing company has spent thousands of dollars to lay out the text, add graphics, and print the book. In short, a number of people thought that this material was important and useful.

This might seem to come off as a major guilt trip, but it's something to keep in mind. Perhaps you might not be learning about a recent, breaking discovery or you might be learning about an area in an introductory course that doesn't grab you as much as other parts of the course. Professors and scientists choose the field that excites them the most, and they, too, are sometimes bored by other aspects of their field. It is completely natural to not find 100% of the material in a course interesting, but as time passes and you establish a foundation of scientific knowledge, you will be able to take more of the courses you like and fewer of those that you don't like. Try to find whatever made scientists before you willing to devote their life's work to this area of study and then let their enthusiasm sustain you.

I Procrastinate Too Much

Perhaps you prefer to surf the Internet or talk to your friends when you really should be studying, or maybe you put off work because you don't want to worry about it. Everyone feels the seductive pull of procrastination, but many successful STEM students have grown to resist these siren calls. Here are some tips to help you get back to work:

1) Some students procrastinate because they can't motivate themselves to initiate the work. If this is you, tell yourself that you are only going to take ten minutes to be completely immersed in whatever you need to do. Chances are, when the ten minutes are up, you'll want to keep working.

2) Make an offer you can't refuse by promising yourself a reward if you complete your task.

3) Get your family or friends to hold you accountable by telling them your work goals. Ask them to check back with you to see whether you've made progress.

4) Visualize what could go wrong if you procrastinate. Think of the potential all-nighter, the anxiety, and the sinking feeling in your stomach as you see your grade. In contrast, think of how amazing you'd feel if you put in enough time and effort to get that A+.

I Can't Focus

Take a step back and figure out what's preventing you from keeping focused. The answer could be as simple as not getting enough water, food, sleep, etc. Even small tweaks to your daily habit could improve your concentration. Your focus could also be affected by a busy schedule. Many students find themselves stretched so thin with friends and extracurricular activities that they don't have the time or the energy for their most important role as students, which is studying! Take a serious look at your planner and narrow your obligations down to the people and activities most important to you. Saying no or turning down opportunities may be difficult, but you'll be happier in the long run and have more time for yourself and your courses.

If you can't focus because you never really had to push yourself in high school, just keep in mind that building up focus takes time and practice. Here we present a modified variation of a study technique once used by Cal Newport— Georgetown University computer science professor and author of the Study Hacks Blog—that you can implement with just a stopwatch and paper.[7]

First and foremost, set a study goal for yourself (finish a problem set, read a textbook chapter, etc.). Then, place yourself in a quiet location and use a stopwatch to measure how long you can study before you start getting distracted. Make sure that the timer is running only when you are focused on your task. When you feel your attention starting to slip away, hit the stop button and record the time you spent studying. Then, take a short break, reset your stopwatch, and start timing yourself again once you get back to work. This time, try to work for an even longer stretch, up to fifty minutes. Imagine that you are training for an academic marathon; think about beating your personal best.

Every night, add up the time you spent studying, record the number on a separate sheet of paper, and do this for three or four days. Timing yourself gives you an objective measure of how much work you've accomplished in a given day. But remember that effective studying is not necessarily about spending lots of time working; more than anything, it is about making sure you are achieving your learning goals.[8] As you continue to challenge yourself, you'll be able to stay focused for a longer period of time and be less susceptible to distractions.

I'm Having a Rough First (or Second) Year Experience
Many students feel a little overwhelmed by college. In fact, it's normal to feel this way for some time. College academics are not like high school. They are harder and cover more material, and your fellow students will generally be brighter and more driven than your peers from high school. Moreover, you're an adult in the eyes of your college. An eighteen-year-old freshman taking a math class with a twenty-one-year-old junior will be held to the same standards by the professor. In short, when you enter college, you've ascended to a new level of education.

To anticipate academic challenges, look over your syllabi, talk to your professors, talk with upperclassmen, and understand how much work will be expected of you. Additionally, know the academic policies at your school, such as whether you can take a course as pass/fail, or the last day that you can drop a course without getting a "Withdraw" or "Incomplete" on your transcript.

Don't overlook the nonacademic challenges you will encounter in college, like living independently and adapting to a new social scene. All of this will take a significant time and effort. You will create new support structures and negotiate day-to-day activities with your roommates. You will feel the pressure to make friends, join groups, and make decisions that place you outside your comfort zone. You will also be responsible for making decisions about eating, finances, health, and managing your time.

Given these stresses, it's not unusual for students to struggle while trying to balance everything on their plates. Your newfound independence in college will be exhilarating, but also exhausting. To prevent new choices from wearing you down, think about what matters to you. How much time is appropriate for partying and how much time should be spent on studying? Would you be willing to pass on a night out with your friends to review for that test? Understanding your personal limits and values will allow you to make choices that you'll be comfortable with.

That said, you may still encounter difficulties catching up with your classes for the first one or two years, and that is perfectly fine. Your courses will likely get more difficult before you find your academic footing, and many students find that their grades in the first couple terms resemble a downward slope. Don't let past disappointments drag you down; learn from your mistakes and concentrate on turning your trajectory upwards by changing your study habits, asking friends for help, talking to your academic advisor, etc. A grade trend that looks like a valley—a decline followed by an incline—will demonstrate your ability to overcome academic hurdles.

I'm Struggling with Something Outside of My Control

Doing poorly on a single important assignment or exam that is worth a lot of points can drag your grade down for the entire term. Sometimes this is due to a simple mistake that could have been easily corrected, like forgetting the deadline for an assignment. Some of your professors may forgive a brief lapse of judgment (which is more likely early in college), but don't count on their mercy. The best thing to do is to learn from the experience and move on.

Your academic course could also be derailed by unforeseeable dilemmas, such as family emergencies, the death of a loved one, or a sudden illness. These scenarios may be unavoidable, and if there is an emergency that might affect how you do in school, reach out to your academic dean and your professors immediately. Professors and administrators know that life doesn't go as planned, and if you talk with them in advance, they will able to work with you to figure out a plan around your current dilemma.

Dealing with tragedy while at college can be particularly challenging if you are far from home, so draw upon the support group that you've formed in college. Universities also have resources, such as professional counselors, to help you cope with the aftermath of tragic events. Your school wants to help you to be well.

I Have That ONE Professor

Not all professors are equal. In fact, some are a lot more equal than others. The quality of your professors can directly influence the quality of your term. If you can choose among different professors for the same course, try to figure out which professors are dedicated and fair lecturers and which are not. Unfortunately, many professors are simply poor instructors. They can enforce an unreasonable grading curve, assign more work than you can finish or too little work for you to learn from, or be unavailable to students. Do your best to research what different professors are like, and seek out those who truly care about teaching undergraduates.

You have many resources at your disposal to gather information about your future courses and professors. Your college may have a system of evaluating its professors that you can use. There are also a number of course-ranking/professor-ranking sites online like ratemyprofessors.com. Keep in mind that the people who bothered to fill out an evaluation online probably had a pretty strong opinion either positive or negative about the course or professor, so more moderate opinions may not be represented. Finally, you can always consult with students who have previously taken the course with the same professor. If you can't get any information on the professor, or a new professor is teaching the course, consider sampling a couple of lectures before the final deadline for adding and dropping a class. Then decide whether or not to continue taking the class.

Some professors are so familiar with a subject that they don't understand how to explain it to a person who doesn't know it already. If you feel like this is happening in a class you are taking, talk to your professor about how you can better prepare for the lectures. Speak with your professor during office hours or schedule an appointment to discuss how you can best understand the material. It's important not to turn the fault on the professors (however guilty they may be).

If your situation in class hasn't improved after attempting to troubleshoot as above, think about taking advantage of the learning materials mentioned in the section on supplemental resources earlier in this chapter. If the teaching you are getting is inadequate, take it upon yourself to find lectures, textbooks, and demonstrations that can give you the information you need. In the worst-case scenario, have a conversation with your academic dean or advisor to consider withdrawing from the course and taking it another term.

I Get Anxious on Tests

All students feel at least some anxiety before taking a quiz or an exam. In fact, some nervousness is completely reasonable and probably even beneficial for test taking. However, test or performance anxiety poses a problem when anxious thoughts overwhelmingly block a student's ability to recall relevant facts or to think through problems logically. Students who suffer from test anxiety may be overcome with feelings of worry, self-deprecating thoughts, and physical symptoms such as tense muscles, sweaty palms, faintness, and nausea.

Your college academic centers will have some resources to help students cope with test anxiety. If you feel that your anxiety is seriously impeding your ability to do well in your courses, bring this to the attention of an academic counselor, who can refer you to the necessary health resources. Your anxiety should be treated as a serious condition, because it is.

Some things will always be stressful regardless of how many times you've done them before. A study of naval pilots during the Vietnam War showed that levels of cortisol—a hormone triggered by stress and anxiety—were just as high for experienced pilots as they were for inexperienced pilots when they landed planes.[9] The experienced pilots didn't get over stress; they simply became more used to it. You have taken many big tests before, that's how you made it to college, but this doesn't mean you shouldn't expect to get nervous before and during the big test—you most likely will. You become better at handling the stress. The key is understanding and accepting this. Even if you feel like you know all the information, it is just as critical to practice taking the test itself, not only to see what the questions will be like, but also to acclimatize yourself to the stressful test environment.

Though coping with your stress is important, there are also steps that can be taken to minimize it. For instance, if you're feeling stressed before or during a test, take 30 seconds and calmly breathe in and out, focusing on the movement of air to and from your lungs. This is a common stress-handling technique, and it will allow you to improve your focus. The key is to bring yourself to the present and focus on what you can do in the moment. On the real test, don't concentrate on the consequences of failure, or even the potential joys of success. Focus on what you are doing, and how you can do it best.

It takes practice and time to get used to handling test stress. If you intend to go to graduate school or professional school, you will have many more years of testing ahead of you. Start learning to handle stress well and to take tests effectively right now.

Conclusion

You now have a sense of what lies ahead in your courses, your labs, your exams, and more. Review these tips and lists, come back to them when you begin your courses, and be flexible enough to try new techniques you may not have explored in high school. Good luck with your first steps into the world of undergraduate science! In the next chapter, we'll discuss different science majors and help you to choose amongst them.

4 Choosing a STEM Major

By the end of your second year or even earlier, you will need to select a major—an area of study in which you will specialize in college. A major will consist of a series of courses intended to give you a deep familiarity with an academic field. Typically, this will be about a third to a half of the courses that you take in college.[1] Successful completion of the major will be one of the requirements for you to obtain your bachelor's degree.

In this chapter, we summarize each of the most general types of science majors:

- Biology
- Chemistry
- Computer science
- Earth sciences
- Engineering
- Mathematics
- Neuroscience
- Physics

Together, the above topics make up the foundation of the STEM fields.

We also discuss some of the considerations and challenges that you should be aware of as you prepare to choose a major. For each major we hope to help you better understand:

- What is this major about?
- What subfields make up this major?
- Which basic courses will I need to take?
- What challenges am I likely to face?
- What can I do to succeed?
- What types of careers can this major lead to?
- Am I likely to be interested in this field?

Even if you have already decided on your major, we encourage you to read these summaries in their entirety—you may discover something new about your intended major and the broader world of science.

Declaring a Major

If you entered college without a clear idea of what you wanted to study, declaring your major can be an arduous task. Look at the descriptions of different majors that we provide in this section and inform yourself about the specific majors available to you. Colleges are divided into distinct major departments that have their own faculty, funding, and concentrations. Your advisors can help you consider majors and courses that you might want to take, but a single advisor is unlikely to have intimate knowledge of every major at your school. To get a balanced perspective about what different majors at your school are like, ask your advisor to refer you to other people that you can talk to, including current students, recent graduates, and professors.

One typical way of selecting a major is to take courses from a wide range of majors during freshman year, contemplate what you liked best, and make a choice with the information at your disposal. Think about the sort of career you want and which concentration will best prepare you for it. Also, take a look at chapter 7, on career choices, before making a final decision about your major.

Many students, some sources say up to 80 percent, will change their major at least once.[2] Frankly, it's best to change your mind earlier rather than later. Some students may find it difficult to finish their major within four years. This is particularly true for students who want to major in more than one subject, study abroad, prepare themselves for specific careers, or apply to graduate school. It can be difficult to schedule all the classes necessary for a major, not to mention any extra courses your college requires you to take. Planning early can make scheduling difficulties much more manageable. It may save you from taking longer to graduate and having to fork over extra tuition. The nitty-gritty information about each major, such as the deadline to select a major, the exact classes you need to take, the professors who teach them, and the times at which they are offered will be provided on your college website. Familiarize yourself with these details so that you can plan out your college career. Spending a couple of hours planning at the beginning of freshman year can save you time, money, and headache later on.

In summary, here are five steps to help you choose your major:

1. Start thinking about academic subjects that interest you.
2. Research major requirements.
3. Start talking to people who are knowledgeable about your major interest.
4. Read chapter 7 about careers, and research which skills and classes would best serve you.
5. Use your school's course list to outline the courses you will need take in each academic term.

To expand on step 5, in addition to your major classes, take note of courses you need to take for graduation requirements, those that interest you on a personal level, courses you need to squeeze in for study abroad and foreign language programs, and prerequisites for graduate and professional programs. Also look for courses that will fulfill more than one requirement and kill two or three birds with one stone. For each term, make sure that you're not overburdened with your course load, and be sure to account for activities that you anticipate will take a significant bite out of your time, such as studying for the GMAT, GRE, LSAT, or MCAT and traveling around the country for graduate school, professional school, and job interviews. Of course, even the best-laid college plans can to go awry. Courses may be canceled, you may decide that what you planned out before may be too difficult to pull off, or your own interests may shift as you progress through college. Be open to the possibility of change, but improvising will be much easier when you have an existing plan that you can tweak. (See Table 4.1.)

Table 4.1. Sample College Course Plan: Pre-Med Student, Major in Biomedical Engineering (BME), with Incoming AP Credits in General Chemistry (§), Intro Physics–Mechanics (§), and Calculus 1 & 2 (§)

Year	Fall	Winter/Spring	Summer
Freshman	**Chemistry 151:** Organic Chemistry I + Lab§* **Engineering 53:** Computational Methods in Engineering* **Math 103:** Intermediate Calculus* **Writing 20:** Academic Writing Seminar‡§	**Chemistry 152:** Organic Chemistry II + Lab§* **Engineering 75:** Mechanics of Solids* **Literature 112:** Modern Chinese Cinema‡§ **Math 107:** Linear Algebra & Differential Equations* *Start looking for research opportunities*	*Summer research*
Sopho-more	**Biology 101:** Molecular Biology§* **Math 108:** Ordinary & Partial Differential Equations* **Mechanical Engineering 83:** Structure & Properties of Solids* **Physics 62:** Introductory Electricity & Magnetism and Optics*	**BME 153:** Biomedical Electronics and Measurements I* **BME 171:** Signals and Systems* **Psych 11:** Introductory Psychology‡§ **Stats 113:** Probability and Statistics§* *Start looking for research and clinical volunteering opportunities*	*Summer research Clinical volunteering*
Junior	**Film 102:** Introduction to Documentary Film‡ **BME 100:** Models of Cellular & Molecular Systems* **BME 154:** Biomedical Electronics and Measurements II* **Psych 103:** Developmental Psychology‡	**Biology 103:** General Microbiology§ **BME 190:** Projects in Biomedical Engineering* **BME 202:** Biomaterials and Biomechanics* **BME 207:** Transport Phenomena: Biological Systems*	*Summer research Clinical volunteering Start studying for MCAT*

(continued)

Table 4.1. *Continued*

Year	Fall	Winter/Spring	Summer
Senior	**Biochemistry 301:** Introductory Biochemistry I§ **BME 462:** Design in the Developing World* **BME 493:** Projects in Biomedical Engineering* **BME 567:** Biosensors* *Start preparing for MCAT* *Start looking for gap year positions*	**BME 494:** Projects in Biomedical Engineering* **BME 590:** Advanced Topics Course* **Public Policy 590-** Science & Tech Law & Policy‡ *Take MCAT* *Prepare medical school applications* *Finalize gap year plan*	*Apply to medical school* *Go abroad* *Start gap year position*
Gap Year	*Gap year job* *Medical school interviews*	*Gap year job* *Medical school interviews*	*Relax, get ready for medical school*

*Major Requirement (or elective within major)
‡Graduation Distribution Requirement
§Pre-Med Requirement

Students Say: What Should Students Know about Choosing a Major?

Study what you actually like and find fascinating. You'll find it much easier to lead a balanced lifestyle if you aren't stuck studying all the time to keep up.

Jeff, University of California, Berkeley

It's important to realize that your undergrad major doesn't lock you into something. It certainly goes a long way in giving you a strong foundation. Much of science is about learning how to ask questions and learning how to approach this kind of field where we're still discovering things every day. I wouldn't feel pigeonholed by whatever is written on your degree.

Grant, Indiana University

Don't think you're stuck in the major that you choose—if you find something else more interesting, go pursue it! Seems like a "duh" thing, but it can be hard to pry yourself away from the beaten path if you've never done it before.

Alvin, Harvard University

Majors, Double Majors, and Minors

During your college education, you will need to complete a major and whatever other graduation requirements are mandated by your college (e.g., earn a total number of course credits, take a particular variety of "general education" courses, obtain a minimum cumulative GPA, etc.). Sometimes, students with multiple academic interests may consider undertaking two or even three majors. Students who want to learn about and demonstrate their expertise in a topic, but cannot or do not want to complete all the courses required for a major, may consider taking a minor. A minor is a secondary area of study that requires fewer courses to complete than a major. Any majors or minors you complete will show up on your college transcript to demonstrate the level of proficiency you have attained in those subjects.

Your school may even allow you to design your own unique major if you can justify why a traditional major would be inadequate for you to pursue your academic interests. This is not a very popular choice for science students. Science classes frequently build on one another and are, therefore, rarely amenable to the mixing and matching of courses involved in a designer major. Nonetheless, if you have a specific idea of what you hope to do, with some careful planning, this may be an option to consider.

Finally, some colleges award honors in a major. Students

usually graduate with honors in their major by completing an honors thesis—an involved course of independent research and a write-up of that research. The honors thesis can demonstrate commitment and depth of experience in an academic field. Note that obtaining honors in a major is different than graduating with honors. Graduating with honors usually means that you had a high cumulative college GPA, and it often comes with a Latin title (e.g., cum laude, magna cum laude, summa cum laude). On the other hand, honors awarded in a major are indicative of your performance only in that major. For further information about honors theses and doing research as an undergraduate, refer to chapter 5.

In general, you can complete any combination of majors and minors in which you are interested. However, at a certain point, you won't have the time or the scheduling flexibility to do any more majors or minors. For this reason, if you want more than one major, do some planning from the start of college. Doing more than a single major is neither worse nor better than concentrating on one subject. If you want to devote more time to learning deeply about one subject or, conversely, want more flexibility in your choice of courses, a single major might be better. If your future career demands that you have credentials in multiple subjects or you simply cannot choose which subject you prefer, a double major or a minor may better suit your needs.

Students Say: Why Did You Decide to Take On a Minor or Another Major?

I originally was planning to do a biology and a chemistry double major. I really enjoyed the chemistry coursework because it was very mentally stimulating. However, I realized that if I dropped my chemistry major to a minor, I would be able to graduate a semester early. I decided to do that because it saved me a semester of tuition, and I got to work in a tissue engineering lab for a semester before starting medical school.

Roger, Wake Forest University

Pursuing a double major will undoubtedly mean sacrificing the flexibility to take classes in a variety of departments. . . . The sacrifice will be worth it if you get to create a unique and synergistic curriculum that makes you passionate and happy.

Sara, Amherst College

I had been interested in mathematics and physics throughout school, and I did not want to give the fields up completely in college. In my mind, secondary majors and minors should be enjoyable. I would not suggest picking a second major or minor for the reasons of padding a CV—such elective tracks should be treated as an opportunity to enrich your college experience and learn something that you really want to know more about.

Peter, Johns Hopkins University, Goldwater Scholar, Rhodes Scholar

Applying for Competitive Majors

At some colleges, certain majors will be so popular that there will not be enough room to accommodate all of the students who want to study them. These selective programs are variously referred to as "competitive," "restricted," or "impacted" majors. Restricted majors will usually require that students complete an application to demonstrate both their interest and aptitude in the field. This application may involve some combination of completed coursework in classes related to the major, SAT/ACT scores, Advanced Placement (AP) or International Baccalaureate (IB) scores, GPA, letters of recommendation, and a personal statement.

If you plan on applying for a restricted major, keep track of the application deadlines and the prerequisites. When the time comes to apply, take your time filling out the application a few weeks before the deadline. If you have to write a personal statement, take a draft to your university's writing center, have a friend read it, and show it to an upperclassman mentor in the major for feedback.

When applying to a highly competitive major, consider whether there is a backup major you would be okay with and whether any of your completed coursework could be applied to it. If you are denied admission to your first choice major, consider your options. Can you reapply or appeal the decision? If so, meet with a professor in the major or another mentor to discuss the steps you could take to improve your application. If this is not possible, a Plan B would be to major in a closely related field and get a minor in your first-choice program. Remember that employers are usually more interested in your specific skills than in what your major was called.

If you can, try to acquire the skills you were hoping to learn from your first choice major by self-study, internships, and carefully chosen coursework. For instance, if you wanted to be a computer science major but were not successful, try learning a programming language on your own, apply to internships at technology companies during the summers, and take whatever computer science classes your schedule can accommodate.

Placing Out of Introductory Classes

At many colleges, incoming students with experience in a certain subject may be able to place out of introductory classes or even gain course credit for their prior work. You may be able to demonstrate this experience by receiving high scores on Advanced Placement/International Baccalaureate exams, taking college/community college classes while still in high school, or passing tests offered by your college to assess new students.

If you can obtain course credit as an incoming freshman, we highly encourage you to do so—this will place you closer to satisfying your graduation requirements. In fact, it may even help you to graduate earlier and save you a lot of money in tuition.

Whether or not you should skip introductory courses during your first year is a more complicated question. On the one hand, placing out allows you to take advanced courses earlier or to take new classes that you may not have had time for otherwise. On the other hand, you may find yourself less prepared than students who took the intro classes. Advanced Placement and International Baccalaureate exams often do not accurately reflect the difficulty of college introductory courses. Moreover, students who elect to take advanced classes in the sciences right away will be competing with upperclassmen who have already taken introductory coursework and acclimated to the world of college-level science. Many science courses are cumulative and sequential; coursework from the past will reappear in more complicated forms and frustrate you if you did not master it earlier.

Even if your college allows you to pass out of a class and pick up course credits, graduate schools and professional schools have specific rules for which credits count toward their prerequisites. For instance, a graduate school might require applicants to take a college statistics course even if they passed out of it through an AP or IB credit. These requirements will be worth looking into later on if you decide you want to go to graduate school or professional school.

In short, past academic experience can put you on the fast track to taking advanced classes and graduating early from college. If you have only a few holes in your knowledge of introductory classes that can be filled with a bit of self-studying, go straight on to advanced courses! If, after consulting with upperclassmen and professors, you feel you need time to adjust to the pace of the college workload, then stick to an introductory course. We recommend taking at least one STEM introductory course—even if you already placed out of it—just to get a feel for the academic expectations at your college. Try to challenge yourself at an optimal level where you are expanding your knowledge and capabilities without being overwhelmed.

Students Say: Should Students Place Out of Introductory Courses?

I wouldn't be so eager to break into upper-level courses in the first year using advanced placement. Consult with a first-year advisor before deciding which pre-matriculation credits or placements you plan to apply, and consider taking department-administered placement tests to more accurately gauge your readiness.

Aaron, Dartmouth College

An Overview of Biology

It seems to me the more we learn about living creatures, especially ourselves, the stranger life becomes.

Lewis Thomas, MD[3]

Biology is the science of life. This science encompasses the molecules that form the basic building blocks of life, the cells that these molecules form, the tissues and organisms that cells organize into, and the ecosystems that exist when different organisms compete and cooperate with one another. Biology encompasses all of this vastness, but it also uses principles about the basic nature of life to organize these topics into understandable systems.

The subfields of biology—many of which may be offered as individual majors at certain schools—include the following:

- Biochemistry: The study of the molecules that make up living organisms and their roles in biological activity, such as metabolism and cellular transport. As an interdisciplinary field, biochemistry overlaps with molecular biology and requires a background in organic chemistry. Courses may be offered by your biology or chemistry department.

- Bioinformatics: The building of computational tools to collect, organize, and make sense of large amounts of biological data. Bioinformatics typically requires a background in computer science and math.

- Cell Biology: The study of the basic units of life, including their variation, functions, parts, and processes.

- Computational Biology: Similar but distinct from bioinformatics, computational biology is the application of mathematical and computational methods to address biological research questions.

- Ecology: The scientific study of the interactions and relationships between living organisms and their environment.

- Evolutionary Biology: The study of the processes that have resulted in biological diversity.

- Genetics/Genomics: Genetics is the study of inherited traits and how genes and their expression vary within organisms and in populations. Genomics takes this a step further and studies how the genome—which includes genes and noncoding DNA sequences—contributes to phenotype.

- Microbiology: The study of microscopic organisms, such as bacteria, viruses, fungi, and some parasites and protists, each of which could be treated as a separate subdiscipline—bacteriology, virology, mycology, etc.

- Marine/Aquatic Biology: The multidisciplinary study of water-dwelling organisms in the context of physiology, behavior, ecology, etc.
- Molecular Biology: The study of the structure and role of proteins, DNA, and other macromolecules important for sustaining biological reactions.
- Neurobiology: The study of the nervous system—the brain, the spinal cord, and the peripheral nerves—at a physiological, cellular, and molecular level. Neurobiology could also be considered a subfield of neuroscience.
- Plant Biology/Botany: The multidisciplinary field of plants and their evolution, development, physiology, heredity, ecological effects, value, etc.
- Physiology: The study of the form and function of organs, cells, and molecular processes in living organisms.
- Zoology: The study of existing or extinct animals in the context of evolution, behavior, physiology, development, ecology, etc.

Basic knowledge of chemistry, math, and physics is required to understand biological processes, and majors will typically be required to take organic chemistry and/or biochemistry.

Coursework taken in the biology department usually begins with a set of introductory classes that survey the various foundations of biology, such as cellular biology, genetics, evolution, and ecology. Many of these courses will involve time working in the laboratory or fieldwork outdoors. From there, you can take intermediate and advanced courses to specialize in one or more biological subdisciplines.

The type of research that this major prepares you to do will depend on the subdiscipline that you choose. If you study molecular biology, then you might find yourself doing research in an air-conditioned laboratory pipetting samples and amplifying DNA sequences. If you are interested in environmental biology, then you may find yourself in the wilderness, fending off bug bites while collecting and cataloging wild specimens. But the boundaries between research modalities are fluid, and research in any given subfield of biology overlaps and interacts with those of other disciplines. For instance, plant biologists played a critical role in advancing genetics. The 1983 Nobel Prize in Physiology or Medicine went to Barbara McClintock, who received her PhD in botany, but later went on to discover genetic transposition.

If your university is affiliated with a medical school or an academic hospital, you may have the opportunity to conduct translational research—research that seeks to use basic scientific principles to understand and improve human health—or clinical research—health research directly involving human subjects. For example, medical research could involve studying the effectiveness of a pharmaceutical compound or the invention of a novel research technique or a medical device.

Successful biology students tend to be comfortable with memorizing and recalling lots of different facts (e.g., "What are the three domains of life?" "Which bacteria can be visualized with a Gram stain?"). After all, life is very complex and is not often describable with a simple equation. Within the diversity of life, there is always some organism or protein that doesn't quite fit into what we understand about the way living things work, and thus simply has to be memorized. On the other hand, *very* successful biology majors are skilled at remembering, but also skilled at organizing what they have learned; taking lots of little details, synthesizing them in a logical manner, and making them into a complete picture.

Career Prospects

Biology majors can use their degrees in a number of jobs, including healthcare consulting and lab/field research. However, because biology is less quantitative than other STEM majors, its application in other fields is limited, and it is becoming increasingly common for biology majors to attend a graduate or professional school for additional training. Biology coursework is one of the main prerequisites for admission to professional school in the health sciences, and biology majors often choose to work in healthcare. In fact, biology is the most common undergraduate major amongst medical students.

Major Advice from Majors

There is not just one prescribed way to be a "biology major." The field of biology is enormous and highly multidisciplinary; it touches upon so many interrelated schools of thought. For example, some of my peers used their talents in computer science to delve into computational biology, or in physics to study microscopy and imaging. Don't be afraid to tailor and individualize your biology major around your personal passions! Once you have a good foundation, take full advantage of your electives to learn about biology applications.

Sara, Amherst College

When professors tell you to memorize the amino acids and such, they mean it. There's definitely a large burden of knowledge, pure factual memorization that you have to do in biology. Some of the mechanisms can also get pretty complicated, so you'll have to have a good mental framework to understand the concepts taught in class.

Jeff, University of California, Berkeley

There is a large quantity of factual information that must be learned in order to understand the field of biology and begin to solve scientific problems. You should dedicate significant time consistently to reviewing factual information covered in class and try to draw connections between different pieces of information.

Ellen, Dartmouth College

An Overview of Chemistry

Double, double toil and trouble;
Fire burn, and cauldron bubble.

The Witches in Shakespeare's *Macbeth*

Chemistry is the science that explains the way in which molecular components form the building blocks of the visible world. By majoring in chemistry, you will develop an understanding of the way that molecular materials interact with one another, how scientists and nature can create and destroy molecular structures, and how to apply newfound laboratory skills to manipulate those structures yourself.

Chemistry is a great major for someone who wants to work not only with problems that have very definite quantitative answers, but also with questions that can be figured out only with a lot of creative thinking or just plain old trial and error. It's a field that stands at the intersection of what we can describe really well with math and what is too complicated or too poorly understood to reduce to a set of equations.

The primary subfields of chemistry include:

- Analytical Chemistry: The identification and measurement of the components that make up a substance.
- Biochemistry: The study of the metabolic reactions that sustain life, and the enzymes that make these chemical reactions possible.
- Computational Chemistry: The use of computers to perform calculations that model real-world chemical systems.
- Inorganic Chemistry: The study of all the elements on the periodic table except carbon, and the compounds that they can form.

- Organic Chemistry: The study of carbon, the element on which all known life is based, and the wide variety of compounds that it can form.
- Physical Chemistry: The application of physics and mathematics to describing and understanding chemical phenomena. Major branches of physical chemistry include thermodynamics, kinetics, and quantum mechanics.

All of these subfields concern atoms and molecules, but each is quite distinct, with regards to both content and the type of thinking required.

The first part of a chemistry major is a series of introductory classes, general chemistry, that will teach fundamental terminology and principles. Basic coursework in calculus and physics will also be required to begin more advanced study of chemistry. After this introduction, a chemistry major will typically study about a year's worth of organic chemistry, a year of physical chemistry, some inorganic chemistry, and some biochemistry. In general, all of these courses will include extensive instruction in the laboratory as well as in the classroom. While analytical and computational chemistry are both important subfields of chemistry, they are not frequently core classes in an undergraduate chemistry major.

Some students may choose to major in a particular area of chemistry. These students will have somewhat different course requirements, but there are usually more shared courses between types of chemistry major than there are different courses. Chemistry majors typically graduate from college with a broad understanding of each of the major subfields of chemistry.

Many of the techniques and procedures you learn to use

as an undergraduate chemistry major are actually used in the laboratory by real scientists every day. Your organic chemistry course will likely involve large, humming machines, intricate systems of piping, and bubbling, colorful liquids in complex glassware—all of which you will employ to synthesize carbon-based molecules. In the biochemistry lab, you will analyze small drops of clear liquid containing the building blocks of life—proteins, DNA, carbohydrates, and lipids—dissolved within them. The lab experiences you get through your coursework will be a strong introduction to the world of scientific research.

Before we finish this description of the chemistry major, we would like to make one important note regarding coursework in organic chemistry. Organic chemistry often gets a bad reputation as a time-consuming and demanding class that discourages aspiring scientists and pre-medical students from continuing to study science. This reputation is not entirely deserved because while it is time consuming and demanding, it is also a very fascinating and enjoyable class as well. Studying organic chemistry is a little bit like playing with Legos or building model cars—it is all about learning how to build molecules. Organic chemistry students often even purchase a toy molecular model set to help them visualize their constructions! While there is a vast body of information to learn about this subject, each piece of knowledge is very much learnable and only requires practice in order to master. Professors, upperclassmen, and graduate students are all there to help you if you take the time to ask them. Moreover, many students find that after becoming accustomed to the rapid pace of study in their organic chemistry class, they are very well prepared to study efficiently for future science coursework.

Career Prospects

When it comes time to graduate, many chemistry majors consider working in the chemical industry or the pharmaceutical industry, going to graduate school, attending medical school, or even working in an unrelated but also highly quantitative field like finance. In addition to improving quantitative skills and reasoning, the chemistry major provides students with knowledge of many laboratory techniques. A major in chemistry demonstrates an ability to study and work hard that will be sought after in many professions.

Major Advice from Majors

The fun and rewarding parts of chemistry involve problem solving and critical thinking, but before you can really dive into that you have to have a solid foundation in the nomenclature and principles of the subject. Learning various named reactions for organic chemistry or figuring out bra-ket notation for physical chemistry can be tedious, but if you learn it well early on you'll be able to get really creative later.

Stan, California Institute of Technology

Repetition is everything. The more times you see a concept on paper, the more times it will stick with you. The nice thing about chemistry is that there is not a lot of rote memorization of facts. Early on, it is important to acquire critical thinking skills to help you learn to solve problems efficiently; work smarter, not harder. . . . Most of the time if I sat down with my book and did many practice problems, my studying was done.

Rachel, Goucher College

Make sure that you're keeping up with class material. Everything in chemistry builds up from previous material, and if you're confused about one area, it's very likely that you won't be able to fully understand newer material. If something in lecture seems confusing, read the book chapter on it, try to do a couple of problems, and attend office hours. I found that for the most difficult material, I needed to have some concepts explained to me, then done for me, then done by myself before I felt comfortable with it.

Daniel, Creighton University

An Overview of Computer Science

Computer science is no more about computers than astronomy is about telescopes.

Edsger W. Dijkstra, attributed

The applications of computer science are an all-encompassing and indispensable part of the modern world. 3D-modeling software, voice recognition programs, smart phones, etc.—these inventions have become the tools that enable modern commerce, manufacturing, and communication. In the past few decades, technology companies like Google and Apple have become giants of the economic landscape.

However, computer science is about a whole lot more than just developing applications for computers. According to Goldwater Scholar and Smith College graduate Emily, "When people think about the computer science major, a lot of them think about gaining competency in multiple programming languages. But computer science is also about understanding computational theories and information systems. These three components are all important and separate areas of the major that will allow you to be a competent computer scientist." Some computer scientists might not even use computers. Such scientists may spend their time trying to answer theoretical questions about how problems can be most efficiently solved or trying to determine what types of problems are solvable.

In this major, you will learn both the theory and application of computer science. Depending on your school, you may need to take courses covering data structures, algorithms, and theoretical foundations. Additionally, you will need to be comfortable with higher-level mathematics, such as calculus, linear algebra, and statistics. Many questions in the sciences and the social sciences can be modeled and analyzed through computational means. If you become proficient in another academic subject in addition to computer science, you may find that you have a unique skill set and can answer important questions about this subject by using your computer science background.

Because computer science is a relatively new discipline, the range of concentrations tends to vary widely depending on the institution and will be in a state of flux for the foreseeable future as the field continues to grow. That being said, some of the most common concentrations in computer science include the following:

- Algorithms: The study of using step-by-step sets of formulas or operations to address problems or perform specific functions.
- Artificial Intelligence: The study of replicating human-like ability for learning, intelligence, and thought in software or machines.
- Computer Architecture: The study of computer design by examining computer components by themselves and in relationship with others. This subdiscipline overlaps with electrical and computer engineering.
- Computational Science: The applications of computer-based tools to solve problems in other STEM disciplines.
- Database Science: The study and development of methods to organize, retrieve, and analyze large quantities of data.
- Digital Arts/Design: The study and creation of digital media, including models and animation.
- Software Engineering: The study and application of software design, development, and evaluation, especially for developing new tools and products.

- Theoretical Computer Science: The study of computational techniques through mathematics.

While grades are an important indicator of ability, real-world experience is probably even more important in computer science. Many of the computer science majors whom we've talked to advise that students who are interested in application-based computer science careers should build an online portfolio of their work to demonstrate their skills to prospective employers. This portfolio can include work completed for a class, but, the most impressive items in a portfolio are often projects that demonstrate creativity and initiative by trying to solve a real-world problem—like a program to help students rate their professors or an application that displays each day's menu at the dining hall on your smart phone. Undertaking a project outside of your coursework allows potential employers to gauge your passion and skills in computer science and gives you an opportunity to expand your knowledge.

In brief, an individual who would thrive in a computer science major is someone who can think through a problem for long periods of time to find methods to solve it logically. Furthermore, many jobs that require a computer science background require collaboration with other people, even with individuals without a computer science background. Therefore, the ability to explain complex concepts is key.

Career Prospects

With all the business applications of modern technology, it shouldn't be surprising that most computer science majors choose to enter private industry.[4] Two common occupations for computer science graduates are software engineering and systems analysis. Software engineers design programs that allow users to run tasks on their computing devices. Systems analysts utilize computer science to analyze a company's needs and to design information systems to address them, thereby allowing organizations to run more effectively.

Internships are critical for providing specific, on-the-job training and for landing competitive positions after graduation. Interviews for internships often involve grilling applicants about a wide range of technical questions. These may involve coding challenges, questions about theoretical computer science, and detailed inquiries about past projects and experiences. Consequently, you should engage in significant preparation and study for an interview.

More than a quarter of recent computer science majors obtain a master's degree after college, usually in the hopes that this will improve their job prospects in industry.[5] In comparison, very few computer science majors elect to pursue a PhD, and those that do are primarily interested in academia. Some topics that computer science graduate students study are data manipulation, machine learning, and knowledge representation. One of the best things about computer science is the versatility of the type of research that you can do both as a PhD candidate or simply as a college student. Many academic disciplines—even humanities disciplines—need computer scientists to help them develop tools to manage and analyze data. Opportunities for research collaborations are widespread.

Major Advice from Students

In an English class, there is a difference between writing an essay and writing a good essay. That's really hard. But when you write a program, it either works or it doesn't work. Then once it works, there's a difference between writing a program that works and writing a good program that works.

Emily, Smith College, Goldwater Scholar

Perhaps the most important study tip for computer science is to stay calm and collected. This is especially true when you're first programming and you encounter a bug. The best way to fix the bug is not to get frustrated, but to ask yourself, "Okay, what do I need to do next?" and to have the expectation that you will be working at it for a while and that this is perfectly fine. For me, doing the homework well meant not worrying when I was stuck on a problem, but instead to keep at it so I could learn as much as possible and develop my problem-solving skills.

Aran, Stanford University, Goldwater Scholar

Knowing whether computer science is right for you is important. It is not for everyone, and forcing yourself to study computer science despite hating it is a recipe for unhappiness. Find out early on whether you are the kind of person who enjoys coding for hours on end. . . . For computational research, look into applied fields and interdisciplinary research and labs with flexibility in methods development. Be proactive in starting new projects and coming up with new ideas. Trying new things loses you very little and often does not take much time in computer science, compared to other scientific disciplines.

Hongyu, Dartmouth College, Goldwater Scholar

An Overview of Earth Sciences

Live in each season as it passes; breathe the air, drink the drink, taste the fruit, and resign yourself to the influence of the Earth.

Henry David Thoreau, *Walden*

The Earth sciences are as varied as Earth itself. Earth scientists seek to answer questions about the formation of planets, the nature of the geologic structures that make up the Earth, the oceans, the atmosphere, and even the origins and evolution of life. They also try to determine how mankind can better coexist with the environment. There is a huge diversity of human knowledge that fits under the Earth science umbrella.

Because of the broad nature of this field, different colleges refer to this major by a number of different names, and even Earth sciences graduates with the same major written on their diplomas may have taken very different coursework. Occasionally this major is called "geology," but this is a little bit of an old-fashioned usage of the word, as geology is technically a subdiscipline of the Earth sciences.

A solid foundation in math, physics, and chemistry is essential to understanding the myriad forces and processes that shape the planet Earth. This means that freshman year for an Earth sciences major might be spent doing a lot of the same introductory courses as many science majors in other subjects. Because of this, if you change your mind later on about Earth science, you will probably still be "on track" to complete another science major. For the same reason, students who have started off studying other sciences may find it to be an easy transition into the Earth sciences major. After finishing prerequisite classes, a much more varied series of coursework begins. Students may go on to learn a little bit about each of the many subdisciplines of Earth science (especially at smaller schools that don't offer super-specialized classes) or they might go on to specialize in a subdiscipline of Earth science while still at the undergraduate level.

The subdisciplines that constitute Earth science include the following:

- Geology: The study of the solid matter that makes up the Earth.
- Geophysics: The interdisciplinary study of the Earth's physical processes, such as volcanoes and earthquakes.
- Geochemistry: The study of the Earth's chemical composition and processes, such as the chemical makeup of rocks.

- Atmospheric Science: The study of the atmosphere, weather, and climate and the processes that influence them.
- Oceanography: An interdisciplinary study of the ocean including fluid dynamics, the composition of seawater in different areas, changes in the ocean on a geological time scale, and mapping sea floors.
- Paleontology: The study of evolutionary and geologic processes through the rock record.
- Hydrology: The study of surface water (rivers) and groundwater.
- Environmental Earth Science: Using Earth science to understand and improve the natural environment.
- Geobiology: The study of the interactions between organisms and the environment and the coevolution of life and Earth.
- Planetary Science: For those who get bored when confined to studying a single planet, planetary science seeks to understand both Earth and other planets (this is also taught in physics departments).

Fieldwork is an important and often truly unique aspect of a major in Earth science. Going out into nature and seeing rock formations, collecting samples, and performing geologic mapping gives students a true understanding of what they have learned in the classroom. Many of the classes within the major will have a fieldwork component to them and, additionally, majors at most schools will be required to go on one multiweek program where they travel to a fascinating natural location (think Death Valley, the Sierra Nevada Mountain Range, or a cruise onto the Pacific Ocean) and hone their Earth sciences skills. Time spent hiking around with professors on these trips can give students a great chance to get to know their teachers in a more personal and casual way than they would in a lecture hall. A love of hiking, rock climbing, and just being "outdoorsy" is common amongst Earth scientists but is by no means required. Needless to say, many students find their experiences in the field to be transformative moments in helping them decide to study Earth science.

In addition to fieldwork, laboratory skills and computer/mathematical modeling skills are quite important in this major, particularly for students who want to go on to graduate school.

Earth may be billions of years old, but in many ways Earth science is a young field that is constantly evolving. It can take all sorts of different backgrounds to tackle a problem the size of a planet. In the words of Harvard geochemistry graduate student Elise, "Earth science tends to be a very friendly, collaborative field because it is so interdisciplinary. That's a big part of the culture."

Career Prospects

The conventional wisdom is that there is a rising demand to protect the environment that is making people with Earth sciences–type skills more valuable to employers. Additionally, the analytical and quantitative skills developed in this major can be broadly applicable to all sorts of careers from law to computer modeling. Nonetheless, jobs that relate directly to Earth science are often limited in number. Earth science students often go on to graduate school, and, in many cases, a master's degree is needed to get a job in an Earth science–related industry like petroleum or mineral exploration. Combining knowledge in Earth science with skills in other areas like public policy can make for unique and hard-to-find skill sets for those who have very specific career aspirations. There are also a few jobs open in academia for students who complete a PhD. Other students who wish to use their undergraduate Earth sciences background in their careers will find jobs in environmental consulting, primary and secondary school teaching, working in labs, working in government (e.g., the National Oceanic and Atmospheric Administration, US Geological Survey, state geological surveys), and working for nonprofits.

With the recent uptick in petroleum exploration resulting from development and use of hydrological fracturing techniques, as well as the graying demographics of petroleum geologists, many of whom were hired in the late 1970s, there are also increasing opportunities for young geologists interested in the oil industry.

Major Advice from Students

Geology is the study of the Earth and Earth history. Geologists use rocks and other natural substances like ice as proxies to reconstruct past states of the Earth and to predict conditions in the future. Geology is a great major if you enjoy traveling and nature. It's a very interdisciplinary major; I enjoyed learning biology, physics, and chemistry and integrating all of this knowledge. Studying and doing homework together is very helpful, since only other geology majors will have the same knowledge base.

Carli, University of Michigan

Doing well in the Earth sciences requires repetition, daily practice, and hands-on experience, spending time with what you are studying and asking lots of questions. Go hiking and enjoy the outdoors while trying to put what you learned in the classroom to use. This is especially relevant with mineralogy, structural geology, and Earth processes.

Lowell, Dartmouth College

Geoscience is a highly integrative, inherently multidisciplinary subject and accordingly it is helpful to reflect on how topics covered in physics, math, chemistry, etc., apply to geology. The nuances are key, but just as important are "big picture" connections. It's easy to memorize equations in physics, but try to connect them to what you're learning about earthquakes, or analyses of climate data sets. I have written off certain topics in other subjects I deemed obscure only to have to call on them in multiple subsequent geology classes. This wide breadth is one of the reasons I remain so enamored with it— biology, ecology, astronomy—it's all there.

Chris, Lafayette College, Goldwater Scholar, Fulbright Scholar

An Overview of Engineering

The optimist says the glass is half full. The pessimist says that it is half empty. The engineer looks at the glass and says: "The space isn't optimized."

Anonymous

Engineering is the major of problem-solving. An engineer is simply someone who applies science to do something useful. Of course, other scientists do useful things as well; the difference is mainly that the focus of engineering is primarily on the *application* of science.

An engineer's efforts to design, create, test, research, and construct practical solutions to a problem can require knowledge from any scientific discipline. Most often, when we think of engineers, we think of people who build buildings and bridges to solve problems of infrastructure. However, an engineer can be someone who works in essentially any field— molecules, computer programs, planes, cars, circuit boards, environmental projects—the list goes on. Because of this wide scope, college students must generally concentrate in a subfield of engineering.

This list breaks down what some different types of engineers do:

- Aerospace engineers design and develop flight vehicles—aircraft, missiles, spacecraft, and satellites—and their various subsystems.

- Biological engineers design, synthesize, and analyze biological systems and molecules.
- Biomedical engineers design and develop biomedical devices, processes, and therapeutic molecules for healthcare. Biomedical engineers usually have knowledge of fields such as biological, electrical, or mechanical engineering.
- Chemical engineers design methods for the production, distribution, and use of chemicals. These products can be found in pharmaceuticals, foods, and a litany of other consumer products. These engineers also ensure that manufacturing processes are sustainable and that production byproducts are sustainably disposed.
- Civil engineers design and maintain public works—roads, bridges, water and energy systems, as well as public facilities like ports, railways, and airports.
- Computer and software engineers design and implement computing systems, computer architecture, software and algorithms, communication networks, and man-machine interface systems. This field integrates electrical engineering and computer science.
- Electrical engineers apply the principles of electricity and electromagnetism to develop electrical and electronic systems; digital computers; power generation and distribution systems; telecommunication systems; control systems; systems for generation, propagation, and reception of radio-frequency signals; signal processing; instrumentation; and microelectronics.
- Environmental engineers maintain and improve environmental quality and optimize the utilization of resources. They design environmental and industrial systems and components, serve as technical advisors in policymaking and legal deliberations, develop management schemes for resources, and provide technical evaluations of systems.

- Industrial engineers design and analyze logistical, resource, and manufacturing systems.

- Materials scientists and engineers design new materials for consumer and industry products, such as computer chips, aircraft hulls, and golf clubs. Some of the materials they work with are metals, ceramics, plastics, and composites. These substances are designed with the goal of meeting certain mechanical, electrical, and chemical requirements.

- Mechanical engineers design products and machines, methods of manufacturing, and systems for energy utilization ranging from automobiles and combustion engines to refrigerators, power plants, and wind turbines.

- Nuclear engineers research and create methods to utilize nuclear energy from fission and fusion of subatomic particles. Applications can include the development of nuclear reactors and medical instrumentation.

- Maritime and ocean engineers design and develop ships, other vessels, and everything from underwater facilities, to oil platforms, to harbor facilities.

- Petroleum and mineral engineers create new methods of extracting oil and gas from deposits below the Earth's surface and discover, extract, and prepare minerals from the Earth's crust.

The subject material you will learn as an engineering major will vary considerably depending on your specialty. For instance, nuclear engineers will take a number of quantum physics courses, which other types of engineers, such as biomedical engineers, may find of limited use in their respective fields. However, all engineering majors will have foundational courses that are integral to the practice of engineering. Training in math and physics is fundamental to engineering, so you can expect to take at least calculus, linear algebra, and differential equations, as well as introductory physics courses in mechanics and electromagnetism. You will probably take a course on systems engineering, which will teach you how to construct and apply mathematical models in engineering. Most students will also need to study some computer science as proficiency with programming is crucial to those in any specialty of engineering. Finally, many majors will take an engineering design class in which they use all of their training to create a product, for example, building a water filter or an artificial heart valve. If the idea of creating something with science appeals to you then engineering may be your field.

It is important to note that some schools, particularly small liberal arts colleges, do not offer a major in engineering. Some of these schools will allow students to apply to finish the last years of their education at a college that does teach engineering so that they can pursue an engineering major.

Career Prospects

Most jobs in engineering involve working within large organizations such as chemical producers, automotive manufacturers, and tech companies. Your role will be to help develop new machines and materials, design systems, and test contraptions—in short, to be part of the team that makes the product, whatever that might be. You will be working to create new things and solve problems no one has ever worked on before.

If you want to get a job right after graduation, carefully choosing your college coursework is vital. Speak to engineers in the workforce to be sure you are taking the right classes to prepare you for your dream job. Specific coursework in a subfield of engineering will make you more employable in that field and save you time with on-the-job training. According to Steven, a petroleum engineer and entrepreneur from Dartmouth, "In engineering, you must design your major around your career rather than trusting that the required classes will have everything you need for the field you want to go into."

Some engineers will decide to pursue a master's degree or PhD. A master's can make you more employable, especially if you felt your undergraduate training was inadequate. A PhD is necessary if you want to become a professor of engineering. In the workplace, a PhD engineer will likely be engaged in research and development, will know how to identify important and challenging problems, and can marshal the resources required to solve them.

Geographic limitations may be particularly important to be aware of for certain engineering professions. For instance, engineers who work in manufacturing will likely find jobs in certain parts of the country that specialize in manufacturing (e.g., Michigan, Indiana, or Ohio). Petroleum engineers will be much more likely to get work in places where there is oil, such as Alaska or Texas. If your dream is to live in New York City, choosing to major in petroleum engineering may pose a bit of a problem.

In addition to more traditional career paths, engineering can open many opportunities for entrepreneurship. As the field is focused on the development of new technologies, engineers at all levels of experience often choose to market their own creations. Some may choose to combine a day job with a side project that they work on during their own time. Those with a passion for entrepreneurship should consider studying the business side of engineering as well, which involves learning how to bring their products to the market and protect their intellectual capital.

Given its focus on problem solving, the career prospects one has as an engineering major are by no means limited to engineering. Indeed, engineering is perhaps one of the most adaptable STEM majors. Fields such as finance, consulting, law, and government are always interested in individuals with engineering backgrounds. Michael Bloomberg, Herbert Hoover, Henry Ford, and Jimmy Carter are all examples of luminaries from all sorts of different areas who received degrees in engineering. In fact, 20% of Fortune 500 CEOs have an engineering degree.[6]

Major Advice from Students

What engineering really does for you is give you a basis for solving any problem. So whether that's directly in your field or not, you have the tools to help you learn what you need to learn and use that knowledge to implement solutions to any problem that you're presented with. And then it allows you to understand when something is not feasible. Because the sooner you can do that, the sooner you can tip it and start doing something more useful.

Matthew, Duke University

What excites me about engineering is the problem solving. It's a process of identifying the problem, defining your design specifications, creating a prototype, testing it, and then repeating the cycle; it's a loop. Essentially, engineering is a really vigorous way of problem solving of defining specifications and having a final product, whether that be an item, service, or system.

Julie Ann, Dartmouth College

The biggest difficulty faced by biomedical engineering students is the diversity of the field and the heterogeneity of programs out there; while some focus entirely on the electrical and mechanical aspects and applications of the field, others focus on tissue engineering or are pre-med focused. If you know what you want to study about biomedical systems or science (e.g., mechanical, chemical, biology), I would recommend choosing those majors as a focus and then taking extra courses in biomedical applications. However, if breadth and not depth (jack of all trades) is what you want out of undergrad, this is the major for you.

Juan, University of Miami, Fulbright Scholar, Goldwater Scholar,

NIH/Oxford Fellow, Gilliam Fellow

An Overview of Mathematics

Pure mathematics is, in its way, the poetry of logical ideas.

Albert Einstein[7]

Math is more than just a type of science; it is a way of thinking abstractly and logically. Essentially, it is about being precise, knowing how you know what you know, and testing your ability to determine how different thoughts and ideas relate to one another. It is the most powerful language we know to describe the natural world, and one of the deepest abstract human endeavors. It is also the framework upon which all the other sciences are built.

There are two main subdisciplines of mathematics: pure math and applied math. Pure math is the study of math for its own sake, particularly the proving of new theorems (true statements), whereas applied math is the application of math to real-world problems. Both involve the creation of new ideas and conjectures, and solving complicated problems. It is the motivation behind the problem solving that usually distinguishes these two subdisciplines rather than the actual methods that they use. Despite this, even the most wildly abstract and seemingly impractical bits of pure mathematics often turn out to have real-world applications that were not initially anticipated—for instance, helping scientists learn how to build a better computer or more fully understand the workings of the atom. Applied math, on the other hand, uses mathematical techniques to model the natural and human worlds, and to solve problems in just about any field from accounting to zoology. Statistics and actuarial/financial mathematics are branches of applied math that some schools may consider to be their own unique majors. Some colleges will offer a single math major while others will offer separate majors in the different subdisciplines of math (e.g., applied math).

Many math departments are flexible in their requirements, but most math programs will require the completion of basic coursework in calculus, analysis (the branch of math that calculus belongs to), differential equations, linear algebra (e.g., matrices), and abstract algebra. Further coursework may vary significantly depending on your interests. Applied math students may be able to fulfill their major requirements with courses in a wide variety of different fields such as computer science, engineering, physics, chemistry, biology, and epidemiology. Math majors who study statistics—the science of describing data and drawing probable conclusions from it—will take additional coursework in statistics, probability, and data analysis. Math majors who study actuarial/financial math will also take unique coursework in economics and finance in order to develop an understanding of financial risk and value.

If two trains are heading towards one another at 50 miles per hour, and a is to b as c is to d, then how many quarters does Sally have? A lot of your high school math coursework may have looked something like this, but college math is quite different from high school math. College math is largely about learning new structures and trying to prove things—figuring out *how* you know that something is mathematically true—and how to make and solve models of the real world with math (sometimes with the help of computer algorithms). This can often require creativity and a lot of patience, but, like anything, you will get better at it with time. Figuring out one problem will set you along a path to figuring out other problems that you didn't even realize were related. You will be stuck 90% of the time, but you will get a real thrill when you solve a problem. The most important thing is that you have persistence so that you can get to a point at which you've solved so many problems that you start to develop a sense of intuition about math and an appreciation for its beauty and power. Try as much as possible to keep track of the overall goal and the meaning behind the problems you are solving. If you run into difficulties, be sure to brainstorm with classmates and ask for help from TAs and professors; everyone finds math to be hard (including professors!), but it can be incredibly rewarding.

Research in mathematics is distinct from the other STEM fields. A research project will usually involve identifying a professor or graduate student who can mentor you, working individually on that research project, and meeting periodically to discuss your progress with your mentor. While other areas of the sciences may require big labs, lots of expensive equipment, and people to run those labs, mathematical research is largely conceptual. For students studying pure math, this can make it even more demanding than other fields because it is hard to pitch in until you have a lot of experience (i.e., you can't be helpful just by cleaning hundreds of test tubes for your professor). The learning curve is steep, and you will need to take many advanced courses before you reach a stage at which you can start trying to answer a research question. In applied math, less training is required to begin doing research. Even if you have only a few classes under your belt, there are so many real-world projects that demand mathematical analysis or modeling that you should have a good chance of finding a research question that you can start contributing toward.

Organizations like the National Science Foundation (NSF) and others sponsor mathematical Research Experiences for Undergraduates (REUs) throughout the country. REUs bring professors together with a group of undergraduates over the summer, giving everyone a forum to work on projects of mutual interest while discussing these projects with one another from time to time. Be sure to ask upperclassmen about their past research experiences and the REUs they have participated in if this is of interest to you.

Career Prospects

Employers in many different fields recognize that mathematics teaches strong problem-solving skills and look very favorably on undergraduates with a background in math. Many math majors say that the skills they developed were directly applicable to their later jobs in other sciences, technology companies, finance, etc. In particular, the ability to make mathematical models that make sense of real-world phenomena is of tremendous usefulness. Other majors stated that, while they do not use their college coursework on a day-to-day basis, it has still benefited them by helping them to think systematically. One particular skill that math majors should consider acquiring before graduating is a basic knowledge of a computer programming language. Knowing some programming can make you more employable and may also help you to apply the math you have already learned to real-world applications or to research.

Typical postgraduate plans for math majors include teaching, finance, consulting, government agencies (e.g., census), graduate school in math, graduate school in other sciences, tech companies, actuarial jobs, and jobs in other industries. If you have an interest in going to graduate school in math or, as many applied math majors often do, going to graduate school in a different field (e.g., biology, computer science, economics), be aware that the courses needed to complete your major are not necessarily sufficient to get you into your desired program. Make sure you are aware of the exact coursework you will be required to finish for your postcollege plans.

Students who are interested in teaching math at the high school, middle school, or elementary school level should consider joining a multiyear teaching program for recent college graduates such as Math for America or Teach for America, or preparing themselves to earn a certification to teach in public schools. We discuss working as a teacher in a bit more detail in chapter 7.

With the rise of "Big Data," knowledge of statistics is in demand more than ever, and there are a growing number of jobs in "data analytics" (e.g., data-mining Facebook posts). A special type of statistician who helps businesses calculate and manage risks (and is often highly paid to do so) is an actuary. Those interested in becoming actuaries must pass a series of rigorous mathematical and financial exams in order to become fully qualified. Interested students should make sure that their coursework is preparing them for these exams and should attempt to start preparing for and taking the exams while they are still undergraduates.

Above all, a career in math can be whatever you make of it. Math applies to everything!

Major Advice from Students

Mathematics is often a very solitary endeavor, and requires a lot of time on your own thinking about it. An important component of math is writing proofs, which is often very different than what students are used to in high school. Adjusting to this can be very difficult, because there is often no "standard" way to go about doing proofs. Just think about the problem, proceed logically, and do not be afraid to ask for help. Everyone finds math to be hard at times.

Carlee, Princeton University

Don't let problem sets distract you from your main goal: getting a general sense of how the math you are learning works, not just solving specific problems. Learning math is like building a castle: you take in all of these ideas and put them together in a way that makes sense to you.

Nilay, Columbia University, Goldwater Scholar

The best students keep track of the big picture and can utilize their problem-solving skills. When you ask them what a proof demonstrates they don't just go through all the steps, they can sum it up. Make sure to learn broadly about different areas of math before choosing one particular field. Applied math majors should take some pure math, and pure math majors should take some applied math.

Joshua, Massachusetts Institute of Technology

An Overview of Neuroscience

If the human brain were so simple that we could understand it, we would be so simple that we couldn't.

Emerson M. Pugh[8]

In 1906, Camillo Golgi and Santiago Ramón y Cajal were jointly awarded the Nobel Prize in Medicine or Physiology for making discoveries that paved the way for modern neuroscience. Over the past century, neuroscience has rapidly evolved into an extremely interdisciplinary field, encompassing neuro-biology, psychology, and some extremely high-tech imaging equipment. Most undergraduates come to college knowing basically nothing about neuroscience, so we are excited to introduce it to you here.

Neuroscience aims to explain how the brain (made up of cells called neurons) works, how we think, and how neurologic diseases manifest themselves. Because many big discoveries continue to be made every year as technology becomes more sophisticated, you will likely learn about many cutting-edge ideas if you choose to study this field.

Because the field of neuroscience is relatively new and because it has some strong overlaps with biology and psychology, not all institutions offer a neuroscience major for undergraduate students. If this happens to be the case but you have a strong desire to major in this field, you might consider seeing whether you can design your own major, as described earlier in this chapter.

Students usually start off this major with an introductory neuroscience course, which covers subjects like learning, memory, sleep, cognition, and behavioral neuroscience. Some schools may also require that majors take introductory courses in psychology and/or computer science as well. A familiarity with computer science will provide you with basic coding and analysis skills for modeling neurons and neural networks.

Neuroscience students often have a huge amount of discretion to take the classes that interest them most—rather than needing to take a specific sequence of classes as in other majors. Every neuroscience major will take courses in the three broad categories described below, but it is up to you to elect which specific courses you'd like to take. The three broad subject areas you will study are the following:

- Cellular and Molecular Neuroscience: The study of the structure and function of the neuron. Coursework in this area introduces students to the action potential, neurotransmitters, receptors, transport of vesicles down the axon of a neuron, etc. It is essentially the cell biology and biochemistry of neurons.
- Systems Neuroscience: The study of how neurons work together to create a brain and how large groups of neurons can enable functions such as language, motor function, and sensation.
- Cognitive Neuroscience: The study of consciousness and other mental functions like memory, and how our nervous system handles these higher-level functions. Cognitive neuroscience exists at the intersection of psychology and biology, and utilizes complicated technology like resting-state, functional magnetic resonance imaging (fMRI), a technique that can monitor activity inside the brain.

Some of the many types of courses that are typically offered by neuroscience departments are Social Psychology, Cognitive Neuroscience, Psychology and Business, Exotic Sensory Systems, Neurobiology of Learning and Memory, Motivation/ Drugs/Addiction, Health Psychology, Principles of Human Brain Mapping with fMRI, Neuroeconomics, Perception, Learning, Cognition, Neuroscience of Mental Illness, Cellular and Molecular Neuroscience, Consciousness, Attention, Systems Neuroscience, Neuroanatomy, Development, Behavioral Neuroscience, Abnormal Psychology, Sleep and Sleep Disorders, and Computational Neuroscience.

You can expect a fair amount of memorization in this field, especially in systems neuroscience courses. The biggest challenge to expect from a neuroscience major is the large amount of information in each course. More than anything else, handling this volume will be a test of your study techniques and diligence. If you approach the material systematically and find out what works well early on, you will succeed in neuroscience.

Career Prospects

According to the Society for Neuroscience, the most common paths for neuroscience majors are academic research, academic administration, pharmaceutical research, work with government-supported scientific initiatives (e.g., the National Institute for Neurological Disorders and Stroke), science writing, and teaching/education.[9] Neuroscience is an extremely active field of research with many unanswered questions. Those who wish to participate in this research at an advanced level may consider going to graduate school in neuroscience (see the section of chapter 7 about going to graduate school).

If the psychological and physical illnesses of the nervous system interest you, you might consider pursuing one of the many healthcare careers that treats and evaluates patients with such illnesses. If you go to medical school after college, you can become a neurologist, neurosurgeon, or psychiatrist and treat diseases like epilepsy, brain tumors, ADHD, and depression (see the section of chapter 7 about careers in medicine and healthcare). Alternatively, you could exclusively treat disorders of mental health by going to graduate school and becoming a psychologist or neuropsychologist.

Major Advice from Majors

Don't cram before exams. Many science courses, including those in neuroscience, require you to understand detailed molecular signaling, enzyme pathways, and physiological processes that are best remembered with time and practice. Also, avoid falling behind on lectures. I liked to use the textbooks for courses to supplement the lecture slides provided and to fill me in on topics I didn't quite understand.

Sydney, University of Michigan

I started college planning to major in biology, but after taking my first neuroscience class, I became very interested in how parts of the brain impact our behaviors in life. I chose neuroscience because it is a lot more applicable to our daily life, which I thought was really cool. I knew after taking my first [neuroscience] course that I had to learn more, and that this major was right for me. Our neuroscience program was very integrated with psychology, which provided a great mix of coursework that I wouldn't have gotten in a biology major.

Angela, Dartmouth College

Many courses, specifically the ones in neuroscience, are very detail oriented while simultaneously big-picture focused. I chose to study neuroscience because I get the molecular detail and precision of neurobiology and the social science perspective of psychology. I liked understanding how and why people think the way they do. When studying, it works well to understand which realm of the major you're studying for and target your efforts as such. Typically, the biological classes require more time, so one must plan accordingly and in advance.

Ameen, University of Alabama, Goldwater Scholar, Rhodes Scholar

An Overview of Physics

Not only is the universe stranger than we imagine—it is stranger than we can imagine.

J. B. S. Haldane

Physics is the study of matter, energy, the forces of nature, and how they work together to make up the universe. Naturally, our understanding of physics influences nearly every facet of scientific knowledge from organic chemistry to cell biology. Furthermore, physics permeates our daily lives. Your computer? It wouldn't have been possible without an understanding of electromagnetism. Your car? An advanced understanding of friction and mechanics made it possible. Certainly, other sciences are involved in these accomplishments—we thank our engineers, not our physicists, for our cars and computers. But physics nonetheless forms the space in which these other sciences operate, the language in which the complex ideas of other fields are communicated.

Core coursework in physics is divided into three main topics:

- Classical (and relativistic) mechanics: The study of the equations and laws that govern how objects behave both under normal conditions and under extreme conditions, such as at light speed.
- Electromagnetism: The study of the various types of electromagnetic radiation that pervade the universe.
- Quantum mechanics: The field that describes of the behavior of molecules, atoms, and subatomic particles.

As a physics major, you will repeatedly visit these topics at increasingly greater levels of detail and rigor. At advanced levels of understanding, the barriers between each of these subfields fade away as you attain a more fundamental understanding of the physical laws of the universe. As you progress to higher-level coursework, physics classes will begin to incorporate even more rigorous mathematics, including techniques from multivariate calculus and differential equations courses. These advanced courses will bring you closer to the important questions currently baffling the minds of physicists across the world.

Should you be given the option of choosing a "concentration" in your major, you may choose to take more advanced classes in a particular discipline of physics (perhaps even at a graduate level), or even try out electives ranging from plasma physics and quantum computing to the physics of early Earth and biological physics.

As you might imagine, the physics major covers some topics that are very complex both conceptually and mathematically. A substantial amount of your homework will likely involve completing problem sets, which require that you really think through every aspect of the concepts you study in class. Physics majors tend to be a tight-knit group of students who work together to solve their problem sets. Group work also prepares students for the collaborative nature of the science of physics.

As in any STEM field, there are research opportunities for physics majors within the physics department. There are also a number of special opportunities for physics research outside of college. For instance, options you might consider include the Max Planck Institute's summer program for undergraduates in Germany or NASA's Student Researchers program. Other opportunities can be found online at the American Physical Society website, which lists a number of programs for undergraduates. In chapter 5, you'll learn much more about opportunities like these, and how to seek funding for research.

Unlike many other STEM fields such as biology, physics has a uniquely strong division between its theoretical and experimental components. For instance, in biology, one laboratory might develop a hypothesis, create an experiment to test the hypothesis, and then interpret experimental results. However, in physics this process is divided. Theorists develop the ideas and experimentalists test them. The reason for this is debatable. One could argue the theory of physics is sufficiently complex that it requires specialization and vice versa for experimentalists. In either case, one half is meaningless without the other. Theory without evidence means nothing and experimental results without analysis is pointless. Given this division, physics majors begin to differentiate into either theoretical or experimental sides of physics while in college. Your physics departments will likely have separate tracks or at least courses designed for students interested in one half of the field or the other. For instance, students interested in experimentation might be able to take courses with more complicated lab sections while students interested in theory could take courses that provide more involved problem sets. Furthermore, if students pursue research, they can work with either a theoretical or an experimental physicist.

Career Prospects

Employers recognize the rigor of the physics major. Like other quantitatively driven STEM majors such as engineering and chemistry, physics majors are sought after in a number of fields, especially those that require complex conceptual and mathematical analysis. Thus, finance and consulting are common non-STEM career paths for successful physics majors. Other popular fields include business, teaching, science journalism, law, and medicine.

Further training is required in the form of a PhD or master's program for a career in academia or government-funded research. A graduate degree will also allow you to work in research and development for physics-related projects at engineering companies. These can range from military defense projects to NASA to medical device development. Additionally, medical imaging and therapeutic radiation delivery require the expertise of medical physicists and dosimetrists, both of which have unique postgraduate training programs leading to DMP or CMD degrees; physics PhDs also play an important role in this area.

Major Advice from Majors

Practice, practice, practice. It is hard to learn physics without solving problems beyond the examples done in class. Try to develop the habit of reading the textbook before lectures to maximize your engagement with the professor or lecturer. Attending office hours is a great way to test your level of preparation; a prepared mind tends to have more insightful questions about physics problems and concepts.

George, Princeton University

The way that physics courses normally go is that you have problem sets at regular intervals. This means that you have to constantly keep up with the reading, and constantly keep up with your work. People who pursue physics are very self-driven. They are the people who think "I have to learn this concept, let me go over it again and again until I get it." As with all other majors, the most successful students in physics majors are the ones who are determined, try hard, and work hard.

Kelvin, Rutgers University, Goldwater Scholar, Marshall Scholar

Conclusions

We hope that these descriptions of different fields will be helpful and illuminating in your initial dive into college science, but no single chapter can ever do justice to the complexity of each scientific discipline. As you go through college and learn more about your likes and dislikes, revisit these major summaries to help guide you in your eventual choice.

5 Conducting Scientific Research

Scientific Progress Goes "Boink"
Title of a *Calvin and Hobbes* Anthology

Open your high school physics textbook. In it, there's certainly some section on Albert Einstein and his theory of relativity. It's an amazing concept that completely changed the way people understood the physical world. It's also pretty new considering how long people have been wondering about how the world works. Only a century ago, even the most educated people on Earth would not have known much of the information taught in today's average introductory science class.

College lasts only about four years, so from a student's standpoint, science seems deceptively static. The fourth edition of your textbook probably teaches the same general concepts covered in the third, second, and first editions. But to researchers, science is ever changing. There are no textbooks. To them, science is a string of open letters that scientists write one another, informing each other of what they have learned or discovered. It's a continual, collaborative effort to build up humanity's knowledge—a library of collective understanding. Scientific research is the process of contributing to that understanding.

Our reality is governed by physical laws, and scientific research is the process of trying to make sense of it all—prodding it, observing it, distilling it, and parceling it up into new forms of knowledge. When researchers attempt to figure out something that no one else knows, they find themselves in wild and uncharted territory. There is no map that can tell them where they are or what to expect (Will this experiment work? Am I using the right conditions? Is this proof correct? Am I even asking the right question?). To understand their surroundings, scientists must continually communicate their discoveries to one another. They do this by publishing their research in academic journals—sharing it with the world so that other researchers can use it.

Scientists publish their research in order to share their findings, but they also publish to demonstrate the value they have added to their field. Each publication, like currency, is evidence of the researcher's success in the business of creating new knowledge. Researchers can then capitalize on their publications by using them to apply for grants, get better jobs, and fund bigger labs, all of which lead to more publications—the cycle of research.

In this chapter you will learn about scientific research, how you can find research opportunities, and how to succeed in doing research as an undergraduate.

Why Do Research?

Doing research gives you the opportunity to join a community of scholars as they push the known boundaries of human knowledge. If the love of science itself is not enough to convince you to dive into research, here are four practical reasons that you should consider.

1) You Will Develop a Deeper Understanding of Your Field

Through research, you will be able to apply the concepts from lectures to address real scientific questions, thereby reinforcing what you've learned in class. You'll also push your understanding of your discipline to a new level when you use these concepts to solve real problems that need an answer.

2) You Will Demonstrate Skills and Perseverance

A productive research experience communicates your intellectual curiosity, capacity to carry out long-term projects, and experience working with others to potential employers and professional schools (e.g., medical school, business school). Furthermore, if you are interested in moving on to graduate school for a PhD or a research-based master's program, research experience is a "need to have" rather than just a "nice to have," as it is *the* most important thing you can do for your application to prove your ability to finish a graduate-level thesis project. (For more information about graduate programs, refer to chapter 7.)

3) You Will Help Generate New Knowledge

Whether you are contributing to a high-profile clinical trial or a small report about a little-known molecular pathway, research will give you the opportunity to bring new knowledge into the world. This, in turn, will help you to appreciate what has already been discovered. In the words of one student whom we interviewed, George, a physics major from Princeton University, the experience of doing research "makes one more than a mere recipient of knowledge."

4) You Might Like It

Plenty of people find the process of doing research, learning laboratory/computational techniques, and becoming an expert in something most people don't know about, to be a lot of fun. By working alongside scientists and graduate students, you will also learn what a career in the area of your research might be like and whether it is right for you.

Students Say: Why Should Students Do Research?

First, research allows student to consolidate information learned in class and teaches them how to apply that knowledge. It also offers a rare opportunity in which students can combine multiple interests, from biomedical science, math and engineering, to social science, into a single, interdisciplinary project. In addition, for students who are unsure about what they want to do, doing research will help inform their career decision and establish a professional network with other students, postdocs, and professors.

Duy, University of California, Davis

Research was one of my favorite experiences from undergrad. Working in a lab, you see that all the work you put in to understand these systems actually pays off when you can accurately predict the result of a reaction. It's also fun—there's a sense of discovery at every turn.

Dan, Dartmouth College

Mentorship opportunities, critical thinking, and self-exploration were probably the biggest benefits to my life that I obtained from research. With the right laboratory and guidance, one can obtain a whole new worldview from even a summer research project. My freshman summer research experience was valuable not because I have a few extra lines in my résumé but because of a close connection with my [advisor], a better understanding of how research works, and a hunger for more research that has yet to die off.

Hongyu, Dartmouth College, Goldwater Scholar

What Should You Expect in Research?

In order to begin doing research, you'll need to train with a scientist or professor who has an existing research group. There, you will learn about the theories, paradigms, and techniques specific to your discipline. While your science courses may have taught a few basic protocols, original research is where real discoveries are made and it will often require a more advanced skill set. In your research group you will learn how to teach yourself when there are no lectures or textbooks to consult, how experiments are designed and conducted, and how hypotheses are supported by data.

Even with the best guidance a student could ask for, scientific research is challenging. Professional researchers devote many *years* to training in their specialized field. Don't let this intimidate you. Every researcher, even your most decorated and senior professors, had to start somewhere.

No matter what sort of research you decide to undertake— and there are many, many different types—keep in mind that setbacks are a natural part of the scientific process. In many cases, your research will result in frustration—machines will break down, data may be difficult to decipher, the code doesn't compile, the cells on the petri dish die out, etc. Don't get too hung up on the roadblocks. It would be much more surprising if everything worked as initially planned. A single sentence found in your science textbook may have been distilled from a lifetime of work. If you could just look up the answer to what you should be doing differently, then it wouldn't be called research. Relax, keep your head up, and try to learn something regardless of whether or not you get any results. The experience of doing research in and of itself is teaching you skills and training you to be a better scientist.

Students Say: What Was Your Research Experience Like?

I worked in a molecular biology lab that I found through fellow Integrated Science Program students who recommended the [research advisor]. I worked there for about 2.5 years including two full summers where I earned grants to work. I was extremely devoted to the lab, spending 20 to 30 hours there each week. I absolutely loved the opportunity because I was able to interact with many grad students, postdocs, and professors and developed the skills to think like a scientist. I've presented my work in several conferences, wrote a thesis, and was the second author on a *Cell* paper.

Robert, Northwestern University

I worked in a lab during my junior and senior years part-time and full-time in the intervening summer. I had had the professor in a class the previous year and got connected with her that way. I got class credit for my research and got paid over the summer. My work didn't get published, but I had my own project and was paired with a postdoc who helped me (it was a really big lab so the [professor] didn't have a ton of time with her undergraduates). I presented to the lab formally in an annual presentation and informally in our weekly lab meeting—these were somewhat stressful at the time but a great learning experience that pushed me to really understand the material.

Sara, Yale University

I did work regarding biosensors for food safety and bioterrorism from freshman year to my senior year. I was able to get some grants from the Department of Homeland Security and through my school. I got to present some work at the National Center for Food Safety and Protection and was fortunate enough to have some things published.

Michael, Michigan State University

I was so eager to begin research freshman year that I jumped on a project that I thought sounded fascinating without really thinking about the type of mentor I would be working with. I was excited to be given my own project and quite a bit of responsibility. However, the [head] of the lab was so famous that he was never there to talk to, and I didn't have a clear mentor. I scraped by due to some very nice grad students and a senior scientist, but I spent two years on a single project that ended up failing. With better mentorship and guidance, I probably could have finished the same project in one year or less, seen it fail, and then applied what I learned to another project within the same lab. However, because I finally realized that I was learning at such a slow rate, I moved to a different lab altogether. Luckily, I had fantastic mentors after that, but looking back, jumping onto a project for the sake of the project itself, without consideration for the mentorship involved, cost me two years of valuable research time.

Mika, Rice University

Publications: How Scientific Knowledge Is Shared
Publications

New findings would be of little value if they never left the desk or the laboratory from which they were discovered. Researchers need to make their results accessible to others for science to progress. For instance, many researchers in astronomy, computer science, mathematics, physics, quantitative biology, and quantitative finance post their pre-print manuscripts in arXiv, an online depository of scholarly works that allows other scientists to access and build upon their works-in-progress.

Most commonly, however, scientists share their findings by publishing their research as an article in a peer-reviewed journal. A peer-reviewed journal is a periodical that publishes research manuscripts after subjecting them to a thorough screening and quality control process carried out by experts in the subject being studied. The purpose of peer review is not to ensure that everything written in the research paper is 100 percent correct, but, instead, to ensure that the paper provides adequate evidence to support what it asserts.

Professionals whose jobs depend on keeping up with the latest scientific knowledge, such as researchers, mathematicians, physicians, and other scientists, often devote a great deal of their free time to reading these journals. The newest and advanced discoveries in any field are almost never published in textbooks because textbooks tend to publish theories, ideas, and experiments that are already pretty well accepted in the scientific community. While peer-reviewed journals put a lot of effort into ensuring the quality of the work that they publish, they're still a place for much more speculative and adventurous types of thinking than is commonly found in textbooks. The experiments and thoughts in any given edition of a journal will be dismissed or embraced by future scientists depending on how they fit in with the results of future experiments.

Articles published in peer-reviewed journals are not the only method for sharing new research findings. Scientists also share their work by attending academic conferences, giving oral presentations, and participating in poster sessions (basically like science fairs for professionals). We will discuss academic conferences and poster sessions in greater detail later in this chapter.

Peer Review

When a scientist gathers enough data to assert a new scientific statement, she can write up a manuscript about her finding and submit it to a peer-reviewed journal in her field of research in the hope that the editors and the reviewers of the journal will select it for publication. This work may be a description of a laboratory experiment and the results of that experiment, a mathematical proof, an explanation of a new tool (protocol), a condensed summary of earlier publications about a given subject matter (literature review), or a variety of other different novel topics. By submitting a manuscript to an academic journal, the scientist is asking her colleagues to evaluate the research, in the hope that they will find it significant enough to be shared with other members of the scientific community. This is the "peer review" part of peer-reviewed journals. Experts who are working for the journal will review the author's analysis, the experiments that were done, and ways the paper could be improved. In rare cases, the reviewers might agree to allow the scientist to publish the paper as is. Sometimes, the reviewers will suggest additional experiments or modifications to the discussion before accepting it for publication. In other instances, the reviewers may decide that the paper is not up to the standards of the journal and will choose not to publish it at all. The process of peer review takes at least a few months and perhaps some rejections, requiring patience and a thick skin.

Many peer-reviewed journals are selective and will publish only a fraction of the papers they receive—those that are of very high quality and importance. Typically, a paper can be submitted to only one journal at any given time, so it is important to choose a journal that is as highly regarded and widely read as possible, but that is not so hard to get into that it will reject the manuscript. The most elite scientific journals like *Science* and *Nature* are read by many scientists in different fields and accept less than 10 percent of articles submitted.[1] In a similar vein, *Annals of Mathematics*, arguably the most prestigious journal in its titular field, received nine hundred fifteen papers in 2013 and accepted just thirty-seven—an acceptance rate of 4 percent.[2] Now, most other journals are nowhere near as difficult to get into, and with the advent of respected online-only open access journals like the *Public Library of Science*, the venues for sharing scientific discoveries are increasing.

Authorship

With the exception of a few disciplines, scientific research is a group effort. As such, many individuals may contribute to a single publication. In order to assign an appropriate amount of credit to each author, the author's names are often listed on the paper in order of the importance of their contribution, the first author generally being the largest contributor and the person who wrote up most of the manuscript. Each scientific discipline has slightly different conventions for how the authors are ordered. For instance, mathematics papers typically list the authors' names alphabetically. Unfortunately, disagreements about the value of individual contributions often lead to conflict among collaborators. In fact, even at the National Institutes of Health, one of the most common complaints that ombudsmen are called upon to mediate is disputes over authorship.[3] For this reason, coauthors should have a clear idea about how their names will be listed on the final manuscript at the onset of a project in order to avoid unnecessary stress and conflict later on.

As an undergraduate, you will most likely work under the supervision of a principal investigator, a senior researcher, or a graduate student who will have most of the say about divvying up lab responsibilities, overseeing the research, and deciding on the authorship of successful projects. While you won't have much authority at this point in your scientific training, discuss these matters with your supervisor before your project starts.

You wouldn't start working for a company without discussing your salary, and you *shouldn't* start working on another person's research project without raising the subject of possible publications and authorship.

Why Publish?

Why do people bother with the process of publishing in peer-reviewed journals? First of all, these journals are widely read by researchers and provide a major outlet for sharing information with the scientific community. Moreover, publication in a peer-reviewed journal lends a certain degree of credibility to scientific discoveries. Finally, as we mentioned earlier, there are other practical reasons as well. Publications are evidence of a successfully completed research experience and, as such, they look good on your résumé, help improve applications to graduate school and professional school, and add value to your application when interviewing with prospective employers.

For professors and graduate students, publications are especially important. Graduate students need to publish if they want to continue working in academia, medical students need to publish to get into competitive medical specialty training programs (called "residencies"), and professors need to publish in order to advance their careers and get funding for their research.

Professors who frequently publish their results in well-respected journals are often rewarded by their institutions with tenure. Tenure means that an institution has contractually agreed to keep employing a professor unless it has strong grounds for firing him (like if he burns down the campus library). In other words, it is a guarantee of job security and it can also come with other financial benefits. There are basically three main types of professors: those who exist outside of the tenure system (lecturers, adjunct professors, research professors), those who are in the process of trying to get tenure (assistant professors), and those who have tenure already (associate professors, professors, endowed chairs, and professors emeriti). Academics who are in the process of trying to get tenure (i.e., the tenure track) are essentially auditioning for a given period of time by working at an institution and trying to amass as many important publications and as much grant funding as possible. Those who succeed are rewarded with tenure while those who are less fortunate may find themselves looking for another job.

Mark of a Productive Research Experience

We've harped on and on about the importance of publications, but the truth is that getting your name on a peer-reviewed research paper during college is a crapshoot. In fact, most students won't see their names on a manuscript from their research experiences from college.

A publication may be a nice addition to graduate or professional school applications, but the admissions committees at even the most selective programs know how difficult it is to get your name on a manuscript. More often than not, programs want to see you that you are able to explain your research so that you can demonstrate what you learned and that your enthusiasm for research is genuine. Plus, there are other ways to show productivity in the lab besides publications, including posters or a thesis, which will be covered later in this chapter.

Research is an incredible opportunity, but being published is really dependent on the luck of the draw. If you happen to join a lab with a project that just needs a few experiments to complete, you can easily get on a paper, whereas you could get stuck on another project for months, if not years. Yet while the process is based on luck, fortune favors the prepared mind.

All of this information about publications is intended more to keep you aware of the larger process that you will be participating in rather than to give you all the tools you need to do it solo. It will likely take many years of training before you start doing independent research. So, how do you get started with your first research experience?

Your first step will be to find a group conducting research on a topic that interests you. Most undergraduates will work in a group that is based in a physical laboratory, but others, particularly students doing math and computer science research, may end up doing work on paper or on the computer. Before you get into the process of actually finding a specific group, look at the handy guide below to familiarize yourself with the different roles that you will find in a typical group and how these roles interact.

In the remainder of this chapter, we'll give you tips on what you should expect from your research experience, how to find a research project, and what to do in order to get the most out of it.

The Anatomy of a Research Group
The Principal Investigator

The principal investigator (PI) is the leader of the research group at a university, industry, or government research institute. The PI oversees the research done under her lab, advises postdoctoral research fellows and graduate students, and brings in funding for projects and salaries by applying for grants from private and public funding agencies like the National Science Foundation and the National Institutes of Health.[4] The more detailed duties of the PI, as well as your future interactions with her, will vary depending on the research group, which can range from one to dozens of students, researchers, and technicians. In some groups, your PI may directly advise you; in other cases, she may rarely emerge from her office. In the future, you'll probably want a letter of recommendation from your PI in order to validate the quality of research that you've performed for her, so make sure to leave a positive impression.

Postdoctoral Research Fellow

Also called a *postdoc*, the postdoctoral research fellow is a researcher who has finished graduate school, has obtained a PhD, and is currently engaged in additional research training. Postdoctoral fellowships used to be uncommon, but nowadays many newly minted PhDs find it necessary to conduct an additional research fellowship or even two if they want to be competitive for a job in industry or to enter a tenure-track position in academia. The PI most often funds the postdoc's salary and provides him with advice and guidance on his projects. Following the completion of a postdoctoral project, the postdoc will enter the academic market and start his own research group, find a job outside of academia, or find a position in another lab to get extra training and more publications under his belt.

Graduate Students

If your research group is located in an institute that grants advanced degrees (e.g., Master's or PhD), then you will encounter graduate students. Requirements for earning an advanced degree can include research rotations, completion of core and elective courses, passing a qualifying exam, teaching undergraduate students, and the production of original research. Graduate training is frequently very intense and these students can often be found working in the lab at all hours of the night and day. (For more information about applying to graduate programs, read chapter 7.)

Laboratory Manager/Laboratory Technicians

It's common for experiment-based laboratories to employ laboratory technicians or managers in order to help out with preparatory procedures and to see to the financial upkeep of the lab. Technicians usually have a bachelor's degree or a master's degree specific to their field of research.

You, the Undergraduate

Your initial work in a research group will be dependent on the group's needs as well as your past research experience. During your first research experience, you may have to prove your worth in the lab by doing some of the more straightforward and uninteresting tasks. It is likely that you will work directly with a graduate student or postdoc to assist them as they conduct their research. As you begin your experience, keep your eyes and ears open—note the location of important items, become familiar with the most commonly used lab techniques and ideas, and be sure to ask questions! Appreciate that graduate students and postdocs are taking time out of their busy schedules to show you the ropes, so pay close attention to what they teach you. The more you demonstrate your competency to those around you, the more autonomy you will gain.

Finding a Research Opportunity

Getting your foot in the door when looking for a research group can be tough, especially if you are attending an institution with limited opportunities for undergraduate research. However, this isn't to say that you should agree to work in any lab that would be willing to have you. Below, we explain how you can find research opportunities and what you should look for in a research group.

Research Opportunities during the Academic Year

Many universities offer programs that match students with available research positions during the academic year through a formal application process. These programs may offer a small stipend to compensate for the time spent in research as well as opportunities to present and share your work with others. Check with your school's office of undergraduate research or your academic department to find out whether such programs exist at your institution.

If you are looking for a research position on your own, start talking to upperclassmen in your major; graduating seniors may know of some upcoming vacancies. Also, reach out to your professors. Even if they cannot help you themselves, they may be able to refer you to some of their colleagues who are looking for a research assistant.

Lastly, check out the research groups at your school by combing through your university's departmental websites. These webpages will often include brief profiles of group members, vignettes of current research projects, and descriptions of recent publications. If you are looking for your first research experience, it may be more important to get some initial research experience under your belt than to find a project you would want to do long term. In fact, consider doing a project in a field different than your major—majoring in one subject doesn't mean you have to do research in that subject alone.

Summer Research Opportunities

For students interested in research, the summer may be the most productive time of the year. During the summer, you'll have the option of doing research without the demands of academic courses and other extracurricular activities. If the type of research you want to conduct isn't available at your university, you can use this time to find research opportunities in other universities, in government institutions (e.g., NASA, US Department of Energy National Laboratories, or the National Institutes of Health), and in industry (e.g., biotechnology, chemical, or engineering companies). Even if you already have a sweet gig in your university during your academic year, taking the time to conduct research in a related lab—perhaps a lab that frequently collaborates with your home lab—will expose you to new techniques, ideas, and people that will help provide you with a wider knowledge base.

Many institutions offer formal summer programs, including Research Experiences for Undergraduates (REU) and Summer Undergraduate Research Fellowships (SURF). Participants are given a modest stipend to conduct research full-time for 8 to 10 weeks at a university or institution. Some of these programs also organize poster sessions in the end to showcase student projects. Because of their perks, these programs can be competitive. For instance, at the National Institutes of Health Summer Internship Program, which hosts around 1,000 interns every year, the acceptance rate is 15 percent.[5] A successful applicant will usually have good grades from STEM courses, strong letters of recommendation, and, most importantly, prior research experience, which makes research during the academic year all the more important.

The National Science Foundation and the American Association of Medical Colleges maintain detailed lists of full-time summer research programs. Other useful databases include the Rochester Institute of Technology's Co-op/Internship list, updated annually. Furthermore, many research-heavy institutions from the University of Michigan to Harvard offer opportunities for students from other colleges and universities to visit and do research. You may have to do some extra digging in their departmental websites to find how to apply for these programs, which typically have deadlines ranging from October to February. If you are interested in participating in a summer program, complete your application materials early and apply broadly. Wherever you eventually choose to apply, remember: all labs are excited to have good research help!

Reaching Out to Research Groups

After you have located a group that you would like to work with, send the PI a short introductory email. Emphasize your interest by mentioning the group's research interests and how the experience will help you achieve your future goals. Include any relevant coursework or previous lab experience. Provide a curriculum vitae (Latin: "course of life") or "CV," an extended version of your academic résumé, as an attachment (for more info about CVs, refer to chapter 6). Also consider including a scanned copy of your transcript and prepare a couple of relevant references, like a professor who taught a class in which you excelled. A sample letter is included below:

Dear Dr. Frankenstein,
I am a sophomore at Ingolstadt University. After hearing about some of the research that you have conducted in the field of tissue rejuvenation, I became very interested in your lab. I would like to inquire about possibly working in your research group next term.

My educational plan is to gain research experience with faculty members as an undergraduate, attend graduate school, and, ultimately, to become a professor of stem cell biology and regenerative medicine. I am familiar with the laboratory procedures and the technical knowledge taught in the three quarters of biology lab courses I have taken: genetics, cell biology, and physiology.

Please find my CV and transcript attached to this email for your review. Upon your request, I would also be happy to send along recommendations from previous mentors and professors. I look forward to discussing the possibility of participating in your group's research with you.

Thank you for your consideration.

Sincerely,
Igor Borodin

Always be courteous and polite, and don't let rejections or the lack of responses get you down. A research group may be reluctant to invest the time, money, and effort involved in training a student. If you don't receive a reply within a week send another email to make sure the previous one didn't get lost. If you don't get any feedback after that, move on and contact another lab. Also consider contacting two or three PIs at a time to see if one suits your interests more than another, but be courteous to your classmates who are also looking for research positions.

If the PI is interested in having you in her lab, she will invite you to an interview to see if you'd be a good fit for the group. Dress professionally, read over any material sent to you by the PI, and skim through the lab's most recently published papers. You can set yourself apart from the pool of candidates by doing some extra preparation. Be ready to answer the following questions:

- What was your previous research experience? (if any)
- What are your professional goals?
- What do you hope to get from this experience?
- What aspect of research do you find most appealing?
- What do you know about this STEM field specifically?
- What positive qualities will you bring to the research group?

Remember that an interview is a two-way street; this meeting is another opportunity to find out whether or not you would like to work in this research group by asking:

- What will be expected of me?
- What is the minimum time commitment?
- Has the group trained undergraduates before?
- Does the laboratory intend to publish the project?
- What sort of contribution would I need to make for the PI to feel confident about putting my name on the publication?

- How does the laboratory decide the order of authorship on manuscripts?
- At what point in my training will I be able to take on my own, independent lab project?

Also consider whether the lab will be favorable for your scientific growth. Is your PI capable of being a solid mentor? If you are interested in graduate school, will the PI go to bat for you for graduate school admissions? Find out whether the laboratory has trained other undergraduates. If there are other college students already working in the lab, ask them how they balance their academics with research and how much mentorship they have with more senior lab members.

In addition, is the lab collegial? Other lab members are the best source of candid information about what your time would be like in a new research group. Because many PIs are quite busy, it is likely that the vast majority of your time will be spent working under a graduate student or postdoc. Will they treat you like a colleague-in-training, a data-generating machine, or a nuisance?

Finally, whether or not your potential PI has tenure may indicate what it would be like to work in their lab. Untenured professors often feel anxious about productivity, since publication in highly regarded journals is important to their future job security and career prospects. While the demand for publication may create a more high-pressure work environment, it may also provide more opportunities for close mentorship, greater responsibility in the research group, and higher productivity. On the other hand, more established PIs may have the funding to carry out lots of projects and employ an army of graduate students and postdocs. They may also have established scientific reputations in their field that could help you make connections later in your academic career, but the caveat is that the more renowned they are, the less time they may have to spend on you. Regardless, every lab is a bit different, and you should do your own research on the matter before agreeing to do anyone else's. Don't be afraid to turn down an opportunity if you feel that the lab is not for you.

Looking for Research Funding

It may be necessary to gain some experience volunteering at a lab before landing a funded research position. If you are thinking of taking up an unpaid research position, ask your PI whether there is potential for paid work in the future. Even if your PI says no, the technical research skills that you will gain could help you secure a paid gig in another laboratory.

Some students get funding from their college or apply for national grants through professional societies (discussed later this chapter) to pay for the time spent in full- or part-time research. Your PIs may be able to help you identify and apply for external funding sources. Look for the resources that are available in your lab and your university.

Students Say: How Did You Find Your Research Opportunity?

I worked in a cellular biology research lab in college that I found by emailing the head of the department and asking how I could get involved. I had to reach out to about five researchers for every one that answered me. Just keep looking for things that interest you, and opportunities will present themselves.

Roger, Wake Forest University

I found my research experience through a school-wide research apprentice pairing program. It was a great opportunity to learn what research was really like and to really work on those skills that aren't learned in the classroom.

Yingchao, University of California, Berkley

I got into research by asking my peer mentor if there were any professors who wanted undergraduate lab assistants. She referred me to one, I went and talked to him, and he connected me with another lab.

Vanessa, University of California, Riverside

At the end of the spring semester of freshmen year, I went up to my professor of biology and asked if I could work with him in his lab. He set me up with a postdoc, and I developed a strong relationship with both of them.

David, University of Southern California

Inside the Lab

Your research experience will differ depending on what type of lab you find yourself in. If you are doing ecological research, your "laboratory" may include the great outdoors. In genetics research, your experiments will mostly involve lots and lots of microcentrifuge tubes and pipette tips. If you are in a bioinformatics lab, you may need only a computer, access to databases, and advanced statistical software.

Don't worry about starting on a small project. Sigmund Freud's first independent research assignment was to try to figure out where eels' testicles are located inside their bodies—a far cry from the theory of the unconscious that he would develop later! Every journey has a beginning, so just start from somewhere and see where your research discoveries and the discoveries you make about yourself take you.

No single book can describe all of the idiosyncrasies you may encounter in laboratories in a vast array of scientific disciplines. However, in all areas of science, one important point remains the same: scientific progress is a cumulative process that depends on the researcher's ability to collect and share data. As such, your best source of information for the type of research that you will be doing will be the papers published in your particular field.

Get to Know Your Field

Present research builds on past findings, so if you are new to the field, you'll have lots of reading ahead of you in order to understand what you will be doing in your research group. Ask your PI and labmates to send you literature reviews—academic articles that summarize the most important developments about a given topic—and read through some of the lab's more recently published articles.

Unless you already have extensive scientific background in your field, you probably won't understand everything in a scientific publication. That's fine! To start, most journals use a similar organization to help you navigate your way around a manuscript, shown in Table 5.1.[6]

Having some intermediate and advanced STEM courses under your belt will help you comprehend research papers, but your understanding will depend mostly on the effort that you put into it. You will constantly run into terms, concepts, and experiments that are new to you. When reading a manuscript, highlight or underline anything that you don't understand, refer to textbooks and online resources to fill the gaps in your knowledge, and then re-read the manuscript.[7]

If you still can't understand components of the article you are reading (and this is normal!), ask your lab members. The

Table 5.1. Organization of a Scientific Article

Component	Description	What You Should Do
Abstract	An Abstract provides a brief overview of the scientific article, including the research question being addressed, the study's hypothesis, the resulting data, and the authors' explanation of what the findings mean.	Scan through the Abstract to decide whether the paper is relevant to your needs.
Introduction	The Introduction explains the scientific background and the motivation behind the study by referencing past discoveries.	Read the Introduction to understand how this scientific article fits within the context of earlier findings in the field.
Methods and Materials	The Methods and Materials section describes how each experiment was set up, what materials were used, and how the data was measured or analyzed.	Refer to the Methods and Materials if you want to know how an experiment was conducted, either to see whether you agree with their methodology or to do the experiment yourself.
Results	The Results section presents the data through figures—graphs, tables, models, diagrams, etc.— with written descriptions.	Look at the data to see whether the authors' descriptions of the figures make sense to you. Sometimes, you may find that you don't quite agree with their interpretation of results.
Discussion/ Observations	The Discussion/Observations section contains the authors' explanation of how the results address the paper's research question.	Think and decide whether the authors' analysis holds up to scrutiny. Are their conclusions logical? Do they overstate their findings? What more could have been done to strengthen their interpretation?
References	The References section is a list of articles that provide the scientific foundation of the research paper.	Look through the Reference section for other papers that may relate to this scientific article.

understanding that you develop by reading and asking questions will make you a much better scientist in the laboratory and the classroom.

Record Everything

Record everything you do in a laboratory notebook: procedures, materials used, data, calculations, observations, and anything that didn't go as planned. Writing everything down helps you to recall and replicate previous experiments or to pinpoint potential errors. The more detailed your description, the easier this will be. Most scientists keep comprehensive laboratory notebooks and folders to keep track of their experimental projects, which can sometimes last for months, if not years.

While you will pick up some basic techniques in your laboratory courses, most of the protocols for your research (e.g., operating a machine, setting up a chemical reaction) will be learned through hands-on training under the supervision of other group members. Angela, a neuroscience major from Dartmouth College who published two first-author papers from undergraduate research projects, offers this piece of advice: "You have to be willing to learn a lot of skills. Most research groups do work that is pretty unique and idiosyncratic. Just being a good student won't mean that you know how the lab works, and PIs realize that. You have to be willing to ask when you do not understand something, and know that your labmates are there to help you."

Always Consider the Big Picture

Ask yourself, why am I carrying out this particular experiment, calculation, or simulation? Whatever you do in your lab, figure out how your experimental task is relevant to the scientific problem being asked. Focus on the bigger picture and you'll learn how data are interpreted and incorporated into a cohesive story designed to make a scientific point. By learning how a scientific question can be addressed, you are building the skills to initiate the next experiment or to ask the next big question in your research group.

Generate Data

After learning relevant background information and techniques, pursue opportunities to run your own experiments and to contribute data to your research group. Talk to your labmates to see what they are doing, figure out what parts of their hypothesis must still be proven, and draw out a testable model that may generate this data. Share the experimental setup with your lab members to get feedback and look through previously published papers to figure out the best controls, variables, and conditions. Depending on its complexity, your experiment may take anywhere from a couple of minutes to months to set up. Run your experiment as carefully as you can and keep a detailed record. If the experiment doesn't work, you will have to modify your methods or come up with a new direction entirely, which will be easier to do with a thorough documentation of your work.

When you finally get your hands on some meaningful scientific data, work your findings into a figure or table as soon as possible, and write a short description of what the data show and how the figure fits into your project.[8] This way, you won't lose sight of the data that you've collected, and if your PI asks you to make a poster or give a lab presentation, you'll have most of your data nicely prepared and ready to go!

At the very beginning of your research experience, you are

bound to make mistakes. Lots of mistakes. Don't get flustered from setbacks or bottlenecks; rather, look at them as opportunities for improvement and learn from your mistakes. You will be able to do this only with practice.

Be Safe

Use your best judgment to evaluate the safety of your research conditions. Every research group must work according to specific sets of safety regulations. However, for a number of reasons that may include cutting costs, lack of oversight, or simple negligence, some labs may not enforce safety precautions as carefully as they should. Other laboratories do work in full accordance with safety standards but, nonetheless, may conduct experiments or use reagents that are relatively dangerous. Research in labs that work with biohazards, poisons, explosives, radioactivity, etc., will always carry some degree of risk, no matter how well-thought-out the experiment. When starting at or considering working in a new lab, ask current lab members if they think the lab's safety practices are up to snuff. Most likely they will tell you that everything is okay; after all, they chose to work there to begin with. At the end of the day though, the choice about what sort of a lab environment you are comfortable working in is a personal one. If you are uncomfortable with your work environment, the best thing you can do for yourself is to find a more suitable one.

If you have an accident in the lab that results in spillage or bodily harm, immediately notify another member of the laboratory and, most importantly, get help. This is not the time to worry about your PI's disapproval. If you need help, get it immediately.

A Final Note about Working in the Lab

Research is an apprenticeship—you are learning by watching others and gradually doing what they do. But unlike for the graduate students, postdocs, lab technicians, and the PI, research productivity does not directly define your future job prospects or livelihood. Prove to them that they should trust you and that you will take seriously the work they do for a living. Observe carefully, soak up as much information and as many techniques as possible, follow procedures, keep your laboratory space clean and organized, and, above all, ask questions.

It's okay to mess up from time to time—everyone does. Expect constant troubleshooting and tweaking of conditions or methods. Once you can run experiments reliably, then the PI and the other group members will entrust you with more important tasks. When you feel ready, ask your supervisor what else you can do to help out in the lab.

Think positively, learn something new, and have fun.

Sharing Your Research with Others

Writing a Scientific Manuscript

With some luck and hard work, your research might yield data that could be used in a scientific manuscript. Most likely, a more senior member of the lab whom you were working with will write most or all of the manuscript. However, in the rare—and incredible—instance in which you are charged with composing the manuscript yourself, you'll find it useful to get a sense of how it should be done by reading through related papers in your field as well as recent papers published in your lab. Every discipline and journal has its own conventions and formatting requirements that can best be learned by reading published papers.

Before writing your manuscript, figure out the point you are trying to make and whether your data will give you enough evidence to make it. Your PI will be a good judge of whether this is the case. In his book *Building a Successful Career in Scientific Research*, scientist and career columnist Phil Dee suggests a great nonlinear order to writing your paper, which we highly recommend.[9] His basic premise is to start with writing the Methods and Materials section first and then to move on to the other sections of your paper. Make sure that your descriptions of the experiment in the Methods and Materials section is accurate and clear. Your goal should be to write so clearly that reader could conduct the same experiment themselves and get the same results. Next, the Results section should contain a detailed description of the data that you've generated from your experiments. This is not the place to analyze what the data mean in the context of your research question—save that for the Discussion section. Explain your data as clearly as possible and don't make any assumptions about what your reader may know.[10]

Now that you have the Results, you have the material to hammer out the Introduction and the Discussion. For the Introduction, start by explaining the scientific background behind your project. Then describe how your research fits into the big picture of your discipline by framing it within the context of past discoveries. Elaborate on the importance of your research, and then delve into your hypothesis and how you approached it. Finally, for the Discussion section, explain how the findings in your Results support or fail to support your hypothesis. Fight to support the conclusions you have drawn from your research with your evidence but don't exaggerate the importance of your findings.[11]

Writing a paper takes time and effort. Unlike with a research paper for class, you (or your PI) will have to set deadlines as you go. Carve out some time in your schedule to work on your manuscript. Once the first draft of the manuscript is completed, solicit feedback from more experienced members of your lab for advice regarding grammar, style, and clarity. Even if you've poured your heart and soul into your paper, be open to criticism and receptive to feedback.

Getting a scientific manuscript published can be a lengthy process and the difficulty involved in publishing will depend on your mentorship, your field, the journal you are submitting to, and the reviewers who read your work. However, this isn't to say that a few enterprising undergraduates haven't done it before.

Beyond publishing in peer-reviewed journals, there are other means of getting your work into the world, especially if you are still waiting to publish your results. A publication in an academic journal is the ideal, but participating in poster sessions and scientific conferences are good additional ways to share your findings and strengthen your academic credentials.

Participating in Poster Sessions

A scientific poster is one of the most common means of disseminating scientific information before the findings appear in a journal. These sessions resemble the science fairs you may have participated in during middle school and high school— lots of big posters sitting around a room with people milling about. Believe it or not, real scientists do this too. Many undergraduate research programs and institutions hold poster sessions to celebrate and display undergraduate research, which allow students to discuss scientific findings with others and to observe the work of their peers. A poster doesn't need to be based on completed research; in many cases, it is a work-in-progress.

A poster should be logically arranged and contain elements of a research paper in abbreviated form, such as introduction, methods, results, discussion, acknowledgments, and a brief reference section (Figure 5.1). However, it's not necessary to have these components specifically written out as subheadings. Your data could also be divvied up into figures and tables or arranged by the points you are trying to make supported by figures (Results) and a brief explanation of your data (Results and Discussions) (Figure 5.2).

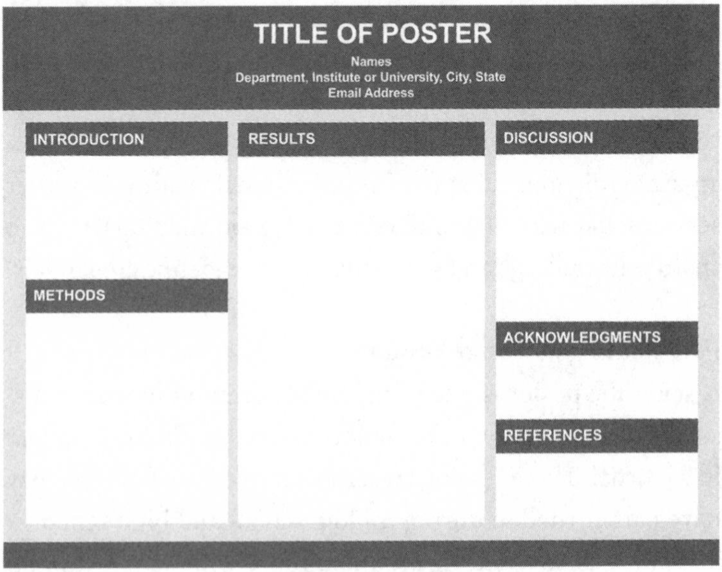

TITLE OF POSTER
Names
Department, Institute or University, City, State
Email Address

INTRODUCTION   RESULTS   DISCUSSION

METHODS

ACKNOWLEDGMENTS

REFERENCES

Fig. 5.1

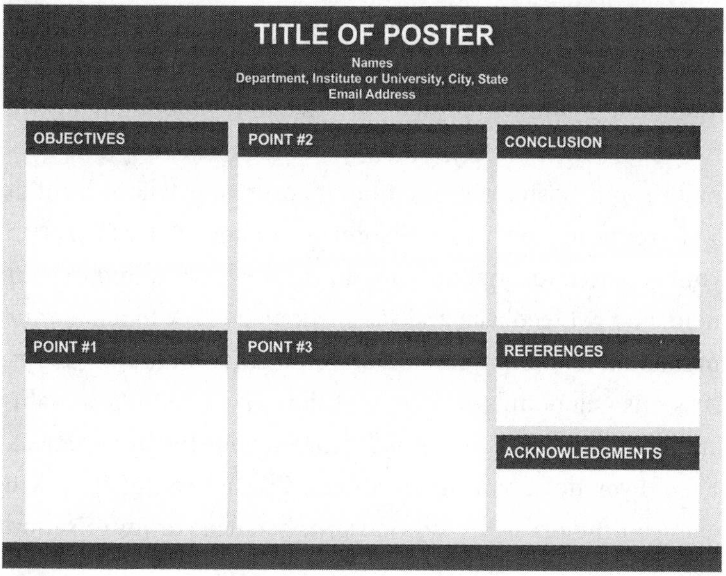

TITLE OF POSTER
Names
Department, Institute or University, City, State
Email Address

OBJECTIVES	POINT #2	CONCLUSION
POINT #1	POINT #3	REFERENCES
		ACKNOWLEDGMENTS

Fig. 5.2

Stylistically, the title should be a prominent part of the poster, and figures and tables should be big enough to read at least five feet away. Be short and concise, and present your information in figures and bullet points rather than long-winded paragraphs. Make sure that your text and figures are well aligned, leave generous margins in your textboxes and in the perimeter of your poster, avoid obnoxious colors, and use a font that is easy on the eyes.[12] A bit of effort here and there will make the poster look organized and professional.

Walk through the hallways of your research building, and you will likely find past posters from your group tacked to the walls. Make a note of the posters you think are the most accessible, and ask your labmates for files of their posters for examples. Typical software used to create posters are Power-Point, Adobe Illustrator, Photoshop, or InDesign.

The information displayed on your poster should be self-explanatory, but prepare a short speech about your research to go along with the poster. Keep it brief—about one to two minutes—and leave the notecard at home. A couple of days before your poster presentation, go over your talk in front of a mirror or in front of your labmates. Better yet, try to explain your research to students who don't have background in your field, and get feedback and suggestions on how to make your presentation more accessible. Some undergraduate poster sessions will double as a competition in which judges evaluate your poster and your explanation of your research project. Even if you don't win a blue ribbon, a poster competition will give you the occasion to hone your science communication skills.

Joining a Professional or Scientific Society

A scientific society is an organization of STEM professionals who work in the same discipline; membership may range from a thousand to over a hundred thousand members. These societies may provide small research grants for scientists, run scientific journals, create networking opportunities, organize scientific conferences, and lobby for increased research funding for their field. That being said, most of the resources offered by societies won't be immediately relevant for you at this point of your scientific training. However, many scientific organizations offer free or discounted student membership rates to get you interested in the field and provide mentorship opportunities, scholarships, and discounts to attend scientific meetings. Table 5.2 lists just a few of the many scientific organizations that offer programs for college STEM students.

Table 5.2. Scientific Organizations That Offer Programs for College STEM Students

Subject	Societies
Biology	American Association of Cancer Research
	American Institute of Biological Sciences
	American Society for Cell Biology
	American Society for Microbiology
	National Association for Biomedical Research
Chemistry	American Chemical Society
Computer Science	Association for Computing Machinery
	Institute of Electrical and Electronics Engineers Computer Society
Engineering	American Institute of Aeronautics and Astronautics
	American Society of Civil Engineers
	American Society of Mechanical Engineers
	American Institute of Chemical Engineers
	Biomedical Engineering Society
	Institute of Electrical and Electronics Engineers
	National Society of Professional Engineers
Geology	American Geosciences Institute
	American Geophysical Union
Mathematics	American Mathematical Society
	Mathematical Association of America
	Society for Industrial and Applied Mathematics
Physics	American Physical Society
Neuroscience	Society for Neuroscience
General	American Association for the Advancement of Sciences

Attending Scientific Conferences and Meetings

The accumulation of scientific knowledge requires the exchange of knowledge from one scientist to another.[13] A lot of this mixing of ideas takes place in scientific conferences and workshops, where researchers present their research, discuss their findings, and network with other researchers. Conferences can be small and organized around a specific subject or quite large with many simultaneous sessions focused on a variety of different topics. The larger conferences are often organized by regional, national, or international societies and held on a regular basis, with attendance ranging from hundreds to thousands of people.

Academic conferences provide opportunities for scientists to communicate their work through poster or oral sessions (i.e., talking about your work to an audience); these oral sessions are reserved for more profound discoveries. Participation in either type of scientific session usually requires the approval of reviewers through "abstract submission." Researchers submit an abstract—a detailed overview of a research project—several months before the conference date. Similar to the peer review process, experts in the field will read the abstracts and select the most relevant and outstanding works to be presented at the conference. The level of selectivity will vary depending on the conference.

If you've made significant progress in your research, talk with your PI about presenting your findings at conferences related to your research. Many major professional conferences, such as the annual American Chemical Society National Meeting & Exposition, offer a separate poster session for undergraduate researchers. There are also a few conferences devoted solely to undergraduate researchers, including the National Conference on Undergraduate Research. If you are thinking about applying to graduate school, a conference can be the ideal location to meet and network with PIs whom you might want to work for in the future.

One of our authors, Justin, makes the following recommendation about conferences:

When I was an undergraduate, I was lucky enough to publish a paper, but when my advisor suggested that I present that paper at a conference, I was reluctant to take time off from class and turned down his offer. After all, I thought, if scientists want to read my little paper, they can always just read it, right? Reflecting upon my choice now, I realize that this was a mistake. Going to conferences can be a good idea for a variety of reasons. At a conference, you can meet people in your field of interest (it's sometimes a smaller world than you might imagine!), get a chance to network, share your results with others, and learn about new cutting edge science. Going to a conference to make a presentation can even be something that you put on your CV. Scholarships may be available from your university or other sources (including the professional society itself) to help you cover the costs of traveling to the conference.

Should I Do an Honors Thesis?

Many STEM major departments offer their students a chance (or require them) to undertake an undergraduate thesis. At many colleges, successfully completing a thesis—in conjunction with an overall strong academic performance in your major—will make you eligible to receive the designation of graduating with honors in your major. Completing a thesis involves the investigation of a scientific problem under the supervision of faculty members followed by a write-up and a public presentation of your research. The entire process will challenge you to reflect on your undergraduate education and may even provide a bridge into your future career. In the best-case scenario, your thesis may lead to a publication or even a place at a scientific conference.

Working on an honors thesis can be demanding and time-consuming. At a minimum, an honors thesis will take at least one academic term to complete. In addition, you may have to "defend" your thesis in front of a small faculty panel, which will involve answering challenging questions about your research to test the level of scholarly development you have attained. The written portion of the thesis is a lengthy work (roughly 20 to 100 pages, double spaced), and will contain sections corresponding to a research paper that would be found in a typical journal in your field. This usually means that your thesis will have an abstract, introduction, methods and materials section, an overview of results, a detailed discussion, and an extensive works cited section, as well as figures, charts, graphs, equations, and tables to document the data you have accumulated during the course of your research.

Consider how important an honors thesis might be to your educational growth or career when choosing whether or not to do one. If you do not intend to make scientific research a significant part of your career or if you want to find a nonscience job after graduation, your time in college may be better spent in other endeavors. Discuss your options with your PI or a major advisor, but keep in mind that they will likely encourage you to do one because it is in their best interest to get more data for their laboratory or for the department.

Awards, Scholarships, and Fellowships for Undergraduate Scientists

A number of awards, scholarships, and fellowships are available to support student research. These will be dependent on your school and project. Some schools may offer research grants and financial support for expenses directly relevant to research, such as purchasing expensive reagents.

If you are a stellar student, you may be eligible for a regional or national award for undergraduates. The most prestigious award available for STEM students with a US citizenship or a permanent residency is the Barry M. Goldwater Scholarship. Each year, the Goldwater Foundation recognizes approximately 250 to 300 sophomores and juniors for high achievement in the sciences—both academically and in research. The scholarship covers some eligible expenses for tuition, fees, books, and room and board and can lead to potential recruiting by graduate schools and industry jobs. For more information about applying to competitive scholarships, read chapter 6.

Be proactive and seek out scholarships, fellowships, and other resources to support your research. As mentioned above, professional societies also offer awards and scholarships for undergraduate students who have excelled in a given area; check out the corresponding national society for your major and/or specific field of research. Don't hesitate to ask your research advisor for a nomination—this is also an opportunity for your advisor to shine as a mentor.

Conclusion

Research will allow you to put your scientific knowledge to use and to learn more about your field of study. Even if you decide not to continue with research after graduation, you will emerge with a better appreciation of how the scientific process is applied in the discovery of new knowledge.

6 Beyond Your Bachelor's Degree

We can never know what to want, because, living only one life, we can neither compare it with our previous lives nor perfect it in our lives to come.

Milan Kundera, *The Unbearable Lightness of Being*

Science can lead you down a lot of different career paths: research, patent law, education, medicine, or even that of an ultra-eligible young author of a book about college science. Even before starting college, you've probably had some thoughts—however vague—about life after graduation. This chapter and the next will help you flesh out those ideas and introduce you to some of the paths often taken by science students. Chapter 6 will help you start thinking about choosing a career, offer a few general tips for entering the working world, and provide information regarding postgraduate scholarships and fellowships. Chapter 7 will talk about the wide array of career and educational options that you have after college, including conventional paths taken by STEM students like graduate schools and less common ones like entrepreneurship. As you continue to develop ideas about possible paths for your future, refer back to these two chapters and the resources they mention in order to learn more about how to prepare for your goals.

But first, here is a giant disclaimer about career advice: There is no formula for your life or your career. Many jobs do not have a standard path that leads directly to them. There is no form to fill out or line to wait in. You have to gradually build your skills and keep your eyes open for opportunities. Perhaps you will even invent your job by identifying and addressing a need that people didn't even know they had.

Our goal for these two chapters is to give you information about postgraduate plans that (1) are relatively common among science students, (2) relate directly to a STEM field, or (3) have standardized application processes. These two chapters should be a good starting point in your thinking about and preparing for the wide world beyond college.

Thinking about Your Postgraduate Life

If you browse through the self-help section of your local bookstore, you'll find two main schools of thought on how you should go about choosing a career. The first school of thought says that you should "follow your passion," and find a job that you love. The other says that passion is irrelevant and that learning how to excel at your job is what will make you love it to begin with, so you should choose something "practical."

If you do, in fact, feel that you have a burning desire in a particular career, be sure that you will be interested in the day-to-day aspects of the job. You may love the idea of being a lawyer and delivering fiery arguments in front of a packed courthouse, but if 90 percent of your time is spent writing briefs, then this may not be the right career move for you. On the other hand, if you don't put much stock in "passion" and see yourself gravitating toward professions with practical benefits—such as a high salary—you should at least choose a career that you can see yourself enjoying over the long term.

Start asking yourself what you want out of your career:

- What sort of work environment do I want?
- What activities will I enjoy working on in the long term?
- Who do I want to spend my time working with?
- What kind of lifestyle do I want?
- How much money would I need to make to support my desired lifestyle?
- Where do I want to live and would this career allow me to live there?
- Am I comfortable with traveling frequently?
- How much job security will I have?
- How much more education and training do I need, and is it worth it?

These are tough questions to mull over, especially when you are already dealing with the everyday demands of college. However, it's best to do some thinking about this in school while you have the time and the resources to do so, rather than waiting until you graduate and have to make a choice immediately about what you will be doing. Many fields require some (or a lot) of preparation during college, and if you can figure out whether you are interested in any of them well before you graduate, you can save yourself from going into a profession you are less interested in due to lack of preparation or putting your life on hold to satisfy requirements you were not aware of earlier.

Fortunately, you don't have to go about it alone.

Talk to College Career Services and the Preprofessional Advising Office

There are professionals out there who are paid to help you choose your future career, get into graduate school, and land a job. These are the staff members at your college's career services department, and you would be wise to seek out their help. They can help you think through the types of jobs you might be interested in, critique your résumé, arrange mock interviews, tell you about jobs you may not have known about, and connect with alumni who have experience in your area of interest.

Many colleges also employ preprofessional advisors who provide guidance specific to graduate and professional school admissions. Preprofessional advisors typically offer targeted workshops, such as preparing for interviews and writing personal statements or statements of interest. These offices can also link you with professionals in your community for job shadowing and informational interviews, and pass on information about summer internships. Get to know your advisors early in your academic career to make sure that you are on the right track to enter the profession you want.

Take a Personality Test

Your career services office can administer and analyze a battery of personality tests that could assist you in determining potential careers.

One such test is the Myers-Briggs Type Indicator, which can tell you about the kind of jobs and responsibilities you may prefer. Another frequently used test is the Strong-Campbell Vocational Interest Inventory, which compares your preferences with those of professionals in a wide range of jobs in order to identify jobs that may be suitable for you. If you disagree with the test results, that's fine; they are meant to help you start thinking about careers, not to shoot down your current goals.

With the results of these tests in hand, talk with staff at career services to find out more about the jobs you were ranked as more likely to enjoy. If you know anyone already working in these fields, meet up with them to ask questions. Career advising offices often have databases of alumni in a wide range of professions. Use these to contact alumni in your area of interest, talk to them about their careers, and begin building professional relationships with them; they may well become your future mentors or employers.

Take a Job for a Test Drive

It's hard to overstate the importance of real-world job experience. You wouldn't commit to buying a car without taking it for a test drive, would you? Try going job shadowing—that is, spending the day following a professional as she goes about her work. Observe the day-to-day happenings on the job—the good and the bad, the exciting and the dull. You can set up a shadowing opportunity by reaching out to your own contacts or by asking your school's career services office to connect you to alumni currently working in your field of interest.

Summer internships provide a more thorough and long-term exposure to careers and may help you get your foot in the door for postgraduate jobs. In fact, some companies extend job offers to their best interns, so think of your internship as a trial period arranged by employers to see whether you'd be a good fit for their company. Even if you don't end up working where you've interned previously, you can still use the experience to make an informed career decision, pick up skills to make you more desirable to other companies, and build networks with others in your field of interest.

Build a Network

Networking is a great way to get insider career information, advice, and opportunities. At this point in your life, your list of contacts likely consists of your friends, your family, and the alumni at your school. Start expanding your contacts by attending social events, job fairs, and professional conventions. You can also grow your contacts through professional social media platforms.

Networking is your best bet to learn about job opportunities before they are posted publicly and, in all likelihood, your best means of actually securing a position. Still, it takes time and effort to seek people out, to ask for feedback and advice, and to convince them to vouch for you. Networking shouldn't end when you get the job you've wanted. Stay in touch with people you meet along the way, and give back by helping those who are still looking for positions.

Here is an example of a letter you might send to a recruitment office at a company you want to work in to learn about a job opportunity and to let your potential employer know that you are interested:

Example 1: Letter for an Informational Interview

Dear Ms. Byrd,

I am currently a rising junior at S College. I first became interested in pursuing a career in investment banking after attending numerous S College Center for Women and Financial Independence events, including your recent talk on Raising Equity Capital.

To prepare for the upcoming recruitment season, I would like to ask you about your company, M, and its recruitment process. I recognize that you may be very busy, and I would appreciate any advice you would be able to offer. Please let me know whether there is a time that best accommodates your schedule.

Thank you for your time and consideration.

Sincerely,

Angie Sun

When networking with someone who is a potential mentor, keep in touch with him or her during your undergraduate experience. Ask thoughtful and earnest questions about the field, such as the following:

- How did you enter your field?
- What are your responsibilities?
- What is a typical workday like?
- What are some of the pros/cons of your job?
- What do people misunderstand about your job?
- What do you think that the job/industry/field will be like in 5, 10 years?

- What are experiences and skills that will make me more market-able in this field?

Take Your Questions Online

You can also find career advice online. For instance, forums and bulletins like Quora, Reddit, Student Doctor Network, or Wall Street Oasis can connect you to a large user base that can answer a wide range of questions specific to your field of interest. Just remember, though, that under the cloak of anonymity, not everyone has the credentials they may claim. Take all online advice with a grain of salt.

Market Yourself

You may not see yourself as a salesperson, but you still need to know how to market yourself. This is true regardless of whether you are looking for a job, applying to graduate school, or applying for a postgraduate scholarship. To this end, you will need to prepare some paperwork with which to introduce yourself to your potential employer, graduate school, or scholarship committee. This paperwork will likely include a cover letter or personal statement, a résumé or curriculum vitae, and letters of reference.

As you begin to look into internships and jobs, find out whether your professional documents require a typical format specific to your field of interest. For instance, a résumé intended for a job in finance is going to emphasize different aspects of your experience compared to a CV aimed at a competitive research program. To figure out the standards for each field, consult your career services office, upperclassmen, professors, and most importantly, those who are already doing what you want to be doing.

Who Are You? The Cover Letter and the Personal Statement

Cover Letter

A cover letter is a professional, one-page introduction to a potential employer. A cover letter should demonstrate your suitability for the position of interest by highlighting relevant technical skills and professional experiences. Your cover letter should answer the following questions:

- How well do your past experiences relate to the position that you are applying for?
- Do you understand the goal of the organization and what it needs?
- How will you be able to address those needs?
- Are you a good fit for the culture of the company?

Frame your cover letter to meet the qualities that the company may be looking for. Look through the company website and analyze their mission statement to see what characteristics they value. Does the company underscore diversity? Explain how your life experiences will help you enhance the workplace. Does the company emphasize commitment to the community? Talk about any volunteer experience you may have and explain how this work is relevant to your application.

Example 2: Cover Letter for
a Summer Job in Business

Dear Mr. Lewis,

I am a junior mathematics major at the University of X interested in working for Company A's Investment Banking division. I believe that my experiences in leadership and quantitative analysis will make me a strong candidate for the division's summer analyst position.

Throughout my academic career, I have explored many opportunities that have allowed me to hone my quantitative skills and leadership abilities. At the University of X, I participated in entrepreneurship competitions in which I analyzed financial statements and market trends and won 2nd place during my freshman year in our school-wide competition. As a former Research Associate Intern at Company B, an international microfinance organization, I wrote financial statements and reports summarizing its long-term returns in emerging countries.

Given Company A's commitment to excellence and to the global community, I am drawn by the company's client-centered philosophy and global market position. I believe that working for Company A will provide me numerous opportunities to create change. Additionally, as an individual who seeks challenges, I believe that Company A will provide excellent opportunities to further nurture my quantitative, analytical, and communication skill set while working with other motivated individuals.

I welcome the opportunity to share with you my qualifications and interest in the Summer Investment Banking analyst program. Thank you for your time and consideration. I look forward to your response.

Sincerely,

Jessica Lim

Example 3: Cover Letter for a Summer Job in a STEM Industry

Dear Ms. Smith,

I would like to request your consideration for the position of project intern for summer 20XX for Company A. I will be graduating from X University next year with a Chemical Engineering major. I believe that developing sustainable alternative energy is a critical scientific issue of our time, and throughout my time in college, I have developed my skills accordingly through academic research and industry work experience.

For three years, I have conducted renewable energy research at X University, including Mobile Biomass Pyrolysis Process Design and CO_2-to-Methanol Process Design. Through these projects, I have gained solid experience in thermodynamics and kinetics modeling.

Additionally I have two summers of work experience in the alternative energy industry as a chemical engineer intern at Company B. At Company B I worked as a Process Engineer to produce cellulosic ethanol from waste biomass. Specifically, I helped to restructure our original protocol to match industry best standards, resulting in a 6 percent increase in yield.

I believe my skills and qualifications provide an excellent match for your position. I would welcome the opportunity to discuss my qualifications further with you or another member of Company A's team.

I will contact you in a week to see whether I may provide additional information regarding my candidacy. If you have any questions, please contact me at ###-###-####.

Thank you for your consideration.

Sincerely,
Christopher Black

Personal Statement

Personal statements are commonly required in graduate school, professional school, and scholarship applications. Different from the cover letter, the personal statement is intended not only to highlight your qualifications and professional experiences, but also to provide a sense of who you are to the reader. Less quantifiable qualities about yourself should be emphasized like how your background and life experiences motivated your decision to apply for the program. The personal statement is also a place to explain how the program you are applying to will help you to meet your aspirations and to convince the reader that you'd be a good fit for the program. Show your enthusiasm for the program by recounting relevant academic and research experiences. You can also include experience outside of your field, such as hobbies and volunteer activities, which portray you as a well-rounded individual. Other points that you may weave into a personal statement are the following:

- What sets you apart from other applicants?
- How have you explored the field (e.g., upper-level courses, research, work experience)?

- What are your professional goals, and how can this program help you get there?
- Have you dealt with difficulties in your life?
- What are some of the personal qualities that you possess that would allow you to succeed in the program?
- How will you advance the mission of the program?

If you are applying to a research-based graduate school program, you may be asked to write a statement of purpose. A statement of purpose is similar to a personal statement, but with a slightly different emphasis. A statement of purpose should be a concise summary of your academic background, an overview of previous research experience, and an explanation of how the program will help you achieve your professional aspirations. Explain what draws you to the specific program by including names of specific professors that you would like to work with in addition to a short description of their current projects. Explain how your past experiences will help you succeed in the program and beyond.

The Résumé and Curriculum Vitae (CV)

A résumé is a concise one- to two-page document that summarizes your education, job experience, and other professional skills. Compared to a résumé, a CV is a more detailed professional summary that contains additional information about academic background, past work experiences, publications, presentations, awards, personal references, and more. Thus, a CV is more relevant when applying to academic, educational, or scientific research positions. Along with the basic components listed above, you could add other relevant details to a résumé or CV, such as career objectives, qualifications, skills

Star Student   Star.Student@uchicago.edu
Cell: (123) 456-7890   123 University Rd. | Chicago, IL 60615

Education

2014-2018   University of Chicago   Hyde Park, IL
                 B.S. Candidate in Mathematics; GPA 3.90/4.00

2010-2014   Smarties High School   Chicago, IL
                 GPA: 3.85/4.00; *Magna cum Laude*

Awards/Honors

2016   Young Investigator's Award
                 Awarded for excellence in a summer mathematics research project at the
                 University of Chicago

Publications

Journal Articles

1. **Student SA**. Why is a baker's dozen equal to thirteen? *Journal of Quantitative Cooking* 1(1),
   100-110 (2016).

Research Experience

Summer 2015   University of Chicago: Department of Mathematics   Hyde Park, IL
                 Advisor: Dr. Cal Culator (Cal.Culator@uchicago.edu)
                 Project: Why is a baker's dozen equal to thirteen?
                 Fellowships Received:
                     National Science Foundation Research Experience for Undergraduates

---

Leadership Experience

2015-Present   University of Chicago Mathematics Society
                 Secretary

2015-Present   Young Entrepreneur Association
                 Chapter President
                 Undergraduate student organization dedicated to helping college students
                 who aspire to be entrepreneurs to invent, develop, and market new
                 products.

Work Experience

2014-2015   Teaching Assistant, Calculus I
                 University of Chicago Department of Mathematics

---

Volunteer Experience

2013-2014   Volunteer Hospital Pianist
                 University of Chicago Medical Center

Hobbies   Violin • Tennis • Journalism

Fig. 6.1

summary, conference participation, a list of relevant courses attended, professional memberships, interests, extracurricular activities, or community service. (See Figure 6.1 for Example 4: Sample Résumé.)

A Quick Note on Grammar

We can't overstate the importance of correct spelling and grammar in your application and follow-up communications with potential employers, graduate schools, and scholarships. Correct grammar demonstrates an attention to detail that is valued in any field. In contrast, typos will raise questions about your professionalism and competency. Have your friends, family, or advisors review your cover letters, personal statements, and résumé to catch grammar or spelling mistakes and ask them how they feel you could improve on your documents.

Many major companies use computer programs to filter through applications, pulling out those with key words and phrases that match job requirements. Avoid unusual fonts and symbols in your résumé/CV and incorporate key words from the job posting in your application material so you can have the best chance of being selected for close review by a human recruiter.

Letters of Reference

Think of letters of reference (also known as letters of recommendation) as a third-party review of your suitability for the program to which you are applying. For this reason, you must request letters of reference from those who are willing to support you without reservation. Ideally, you should ask your potential letter writer in person, but, if that is impossible, it is also acceptable to send an email. Be certain to ask them whether they would be willing to give a strong and positive recommendation. A lukewarm reference will do you more harm than good, and when asked, most individuals will be honest about whether they will be able to vouch for you enthusiastically. You can guide the content of the letter by pointing out the qualities that the letter writer can specifically remark about you better than your other potential recommenders can. Try modeling your request for a letter of recommendation after this template:

Example 5: Request for a Recommendation Letter

Dear Dr. Tran,

I hope that you are doing well.

I am writing to ask whether you would be willing to write a strong letter of recommendation for my application for the National Institutes of Health Postbaccalaureate Intramural Research Trainee Award Program. This opportunity will allow me greater exposure to research as I consider applying to graduate programs next year.

My other references will be able to talk about how I have performed in their courses, but you have seen me in a professional research environment. For this reason, I was hoping that you could talk about my past work ethic in your lab and my interest in research.

For your convenience, I have attached my résumé, my academic transcript, and the personal statement I am using for my application. The letter will be due on May 1 and should be emailed to XYZ@NIH.gov prior to that time. Please let me know if you have any questions.

Thank you for your time. I will look forward to your response.

Sincerely,

Carlos Esperanza

Recommendation letters are written on your behalf, so it is an honor that others are taking time out of their busy schedules to help you achieve your professional goals. Make sure to give your letter writers time to craft their recommendations, four weeks at the very least. According to Dr. Kevin Haworth, an assistant professor at the University of Cincinnati, "After requesting the letter in person, send the recommender all the information they will need in a single email. They will likely need to know what the letter is for, the due date, how to submit the letter, to whom the letter should be addressed, and the address of the person to whom the letter should be addressed." You can provide your letter writers with your résumé, your transcript, an autobiographical sketch, and an explanation of how the opportunity will help you attain your future goals.

If need be, remind your letter writers one or two weeks before letters are due. After the letters are submitted, send a thank you note, and remember to update them on the results and your decision at the conclusion of the application process.

Acing the Interview

If you receive an interview invitation, this is a good indication that you are a competitive candidate for the program or position to which you are applying. The interview is the best opportunity to pitch yourself to potential employers, admissions committees, or scholarship committees, so try to make a strong first impression.

To prepare for an interview, review the details of the position or program for which you're applying. Furthermore, research the company or institution with whom you are interviewing. Think of how you'd like to portray yourself to your interviewer. Some good questions to ask yourself are similar to what you may have written in your cover letters:

- Why do I want this job/scholarship/grad school position?
- Why am I the best candidate for this program? What sets me apart from others?
- What are my strongest assets?
- What are some of my strengths and weaknesses?
- What are my professional goals?
- Where do I see myself in five, ten, or fifteen years?

These are common questions that can be asked at any interview. By preparing your responses ahead of time, you can develop a cohesive message to send to your interviewer.

When applying for a job, remember that employers want intelligent, professional, driven, and well-spoken individuals. Employers also want someone who will be a good team member and will interact well with colleagues, customers, and clients. They will be trying to assess whether you have a good attitude and qualities like empathy and integrity. Since most companies have some type of values description incorporated in their mission and vision statement, you can find out ahead of time, through research, what qualities they look to build in their company. Focus your responses to the above questions to channel these qualities into your answers. Your prospective employer will want to know that you are going to be engaged and add value to their team and customers, solve problems, and be energetic in trying to accomplish the company's goals. Help them to get a sense of what sort of coworker you would be.

Nearing the end of the interview, your interviewers will ask you whether you have any questions. Don't waste time by asking questions that could easily be answered by quickly browsing through the Internet. Instead, come up with thoughtful questions that touch on unique aspects of the program. This will show initiative and preparation on your part and can help distinguish you from other applicants.

After the Interview

At the end of the day, send an email to your interviewers to thank them for their time and consideration. Try to put in a personal touch as well by mentioning something about your interview that you found particularly moving or interesting. No one wants to get a generic thank you letter. You can also send a handwritten note, which provides an additional personal touch that email does not. If you are going to an onsite interview, take some stationery and postage stamps with you, and write a short note to your interviewer and send it off before heading home. This will make sure that that the details of your conversation are fresh in your mind.

Example 6: After-Interview Thank You Letter

Dear Ms. O'Brian,

Thank you for taking the time to meet with me on Monday. I enjoyed our conversation, and I appreciate the University of X's interest in my application. I particularly enjoyed our discussion of your path from working in business to becoming a doctor. I am excited about the possibility of training to become a physician at the University of X and taking advantage of its amazing array of resources and opportunities.

Again, thank you very much for your time and consideration.

Best regards,
Makayla Johnson

Be Professional

Another important aspect of interviewing is to dress and present yourself professionally. This is true regardless of whether you are applying for a job, for graduate school, or for a scholarship. Unless otherwise noted, a job interview will require formal work attire. For gentlemen, this is a suit and tie. For ladies this could be dress pants or a skirt with a formal shirt. Make sure your clothes are ironed and clean prior to the interview. Don't wear too much make-up. Hair should be styled conservatively and professionally.

During the interview day, sit up straight, make appropriate eye contact while speaking, don't touch your face, and remember to smile. When replying to a question, consider your response and speak clearly and deliberately.

Being able to convey professionalism is a necessary work skill and a necessary life skill. Essentially all the rules you need to be professional in the workplace or at an interview were taught to you in high school. Put your cell phone away and on silent (or "Do Not Disturb") to be safe. Don't curse or use foul language. Be friendly to everyone you meet, including the secretaries, the janitorial staff, and your potential coworkers; being friendly will help you feel positive and energized during the interview day, and you don't know who will be asked about your presence and energy after the interview. Have a good attitude. Exercise personal grooming. Use proper grammar. Always, always, always be on time.

While these rules may seem basic, you'd be surprised at how many individuals break them. Employers often see such transgressions as red flags, whether you've already been hired or not. According to Amanda, a Bowdoin College math major and financial analyst at Goldman Sachs, "A lack of professionalism is something that will absolutely prevent you from getting a job."

Build Your Online Presence

In addition to the materials you officially submit to your potential employer, consider setting up an online profile describing your professional projects and interests.

For instance, through LinkedIn you can build a profile that displays your education, past and current employment, relevant and noteworthy experiences, honors and awards, marketable skills, publications, and any other accomplishments. As you build your profile, grow your connections with people and, wherever possible, ask people who know you, your experience, and your work habits, to endorse you for certain skills you have or write an online recommendation for you. Seek out professional organizations and groups in your current and prospective areas of interest and join those groups to build more connections.

Consider creating your own website as a portfolio of your past projects and current interests. You can purchase a domain name through a registrar like GoDaddy.com or set up your own website for free through a hosting service like WordPress.

Additionally, be aware of any unflattering content about you that may exist online. If you can, get rid of any online postings that could undermine your character to an interviewer. According to a 2014 survey conducted by Jobvite—a recruiting management company—a full *93 percent* of recruiters checked out an applicant's social media before making a decision on their candidacy.[1] Fifty-five percent of recruiters reported that they reconsidered their decision based on what they had seen, with 61 percent of these reconsiderations being negative. Some of the most common red flags in the candidate's social media were profanity, incorrect spelling and grammar, references to alcohol and drugs, and sexually explicit posts. Untag yourself from any photos you wouldn't want your future employers to see and ask your friends to take them down.

Postgraduate Grants and Fellowships

There are many grants and fellowships aimed at helping accomplished college graduates continue their studies. These include stipends or tuition for postgraduate programs or projects, such as graduate school, studying abroad, research, or language study. The first place to look for these postgraduate grants is your university. Additionally, some of the best-known—and most competitive—fellowships and grants are described in Table 6.1.

Many competitive national awards will require an official endorsement from your university. Schedule your appointment with your institution's scholarship office early in your college career to discuss your eligibility, your competitiveness, and the preliminary application process specific to your institution.

There are also many smaller scholarships and fellowships, some of which may be specific to your college. These programs may have very narrow specifications for students with a certain racial background, religious faith, geographic location, career interest, and so forth. Start searching for postgraduate fellowships by the summer before your junior year. Talk to career services to see which awards could be relevant to your interests, and consult your professors, upperclassmen, and databases like Scholarship.com to expand your search.

Table 6.1. Postgraduate Fellowships and Grants

Scholarship/Fellowship	Description
Churchill Scholarship	Nine to twelve months of tuition, fees, and living stipend for eligible graduates of participating colleges to study mathematics, physical sciences, biological sciences, or engineering at the University of Cambridge.
Deutscher Akademischer Austauschdienst (DAAD) Study Scholarship	Scholarship and stipend for graduates of North American universities to conduct a research project or complete a postgraduate degree at a German institution.
Gates Cambridge Scholarship	Funding for students from outside the UK to pursue a full-time postgraduate degree in any subject available at the University of Cambridge.
Ford Foundation Fellowship	Three-year fellowships for graduate students from underrepresented groups pursuing a PhD or ScD at US institutions.
Fulbright Scholarship	Teaching or research grants in a variety of fields for two months to a year in over 125 countries.
Gilliam Fellowships for Advanced Study	Up to three years of stipend support for 2nd or 3rd year graduate students from disadvantaged and underrepresented groups in STEM pursuing a PhD in the life sciences.
Hertz Fellowship	Stipend and tuition for students pursuing a PhD in the physical sciences, biological science, and engineering.
Luce Scholars Program	Stipend, language lessons, and professional placement for eligible graduates of participating US universities to live and work in Asia for one year.
Marshall Scholarship	One- or two-year tuition and stipend for US students to study in the UK in a variety of academic disciplines.
Mitchell Scholarship	One academic year of tuition, housing, stipend, and travel for postgraduate study in Ireland or Northern Ireland.
National Science Foundation Graduate Research Fellowship	Funding for research-based master's or doctoral degree in the natural, social, and engineering sciences at US universities.

(continued)

Table 6.1. *Continued*

Scholarship/Fellowship	Description
The Rhodes Scholarship	Funding for non-UK students to study at the University of Oxford's postgraduate programs.
Whitaker International Program	Stipend for biomedical engineers to conduct research, take classes, or take an internship anywhere outside the United States or Canada

How to Be a Competitive Applicant for Undergraduate and Postgraduate Fellowships

In chapter 5, we briefly mentioned some awards that students can apply for, such as the Goldwater Scholarship. Most of these major awards focus on five factors:

- Academic achievement
- Leadership/extracurricular achievement
- Letters of recommendation
- Personal statement
- Future potential

This section is made up of advice from winners of prestigious scholarships and fellowships, but its lessons should be applicable to virtually any award, competitive or otherwise.

Start Early

Plan ahead to meet all the expectations of the program in which you are interested. For example, to be competitive for the Rhodes Scholarship, you not only have to excel academically, but also have "fondness for and success in sports" and "moral force of character and instincts to lead, and to take an interest in one's fellow beings."[2] Demonstrating that you possess these characteristics takes time and effort. Understanding the requirements of a particular fellowship ahead of time will make you a more competitive candidate.

Start Small

It isn't uncommon to see recipients of major grants and fellowships win other prestigious accolades. The more awards you win, the more competitive you will be for other awards. "There are a number of small scholarships and fellowships that you can win at any level [of your college education]. If you can get involved with these as early as possible, you will only be pushing your career forward," says Juan, a recipient of a Goldwater Scholarship, a Fulbright Scholarship, an NIH/ Oxford Fellowship, and a Gilliam Fellowship. "When you get an award, no matter how small, it shows people about your enthusiasm and dedication to the field, and will only help you be more competitive for the next award. No one jumps to a Rhodes or Fulbright. It's a stepping-stone process."

Demonstrate How the Award Fits in the Context of Your Goals

Let's step into the shoes of the selection committee. It wants to select applicants who have the best chance of making a positive change in the world that, in turn, will reflect positively on the organization. If the award is meant to help students pursue a STEM PhD to make an impact in their discipline two or three decades down the road, then the committee probably doesn't want to give the award to someone who will eventually jump ship for, say, corporate finance. You have to highlight previous accomplishments in a way that demonstrates fit for the award that you are applying for. According to Peter, a Goldwater Scholar and a Rhodes Scholar from Johns Hopkins, "One of the most important parts of applying for a scholarship or fellowship is figuring out how to talk about your passions. Whether in written application essays or interviews, you need to develop your own voice to be able to clearly and genuinely discuss the narrative of your life and your goals. Speaking with advisors and mentors is crucial to this process, and a sincere discussion of your passions with someone else can help you figure out exactly how to articulate what makes you tick."

According to Dr. Kevin Haworth, the benefits of winning a recognized award can have a long-term effect on your career. "As a (2001) Goldwater Scholar, I can attest that receiving the Goldwater can be a defining factor on your résumé and may be one of the things that sets you apart from other graduate school applicants. While the monetary award is quite welcome, the recognition that comes with being named a Goldwater Scholar is more than worth its weight in gold."

Apply! Apply! Apply!

You will never win a scholarship if you never apply. Make time in your schedule to look for scholarship opportunities and fill out application forms. According to Kelvin, a Goldwater Scholar and a Gates Cambridge Scholar, "I was very fortunate to have a scholarship advisor who told me about these opportunities, and there really are so many scholarships out there. Ultimately, the worst thing that can happen is that the application committee will say no, and the best thing that can happen is that the opportunity will completely change your life. So, take a chance and apply."

Students Say: How Can STEM Students Best Prepare Themselves for Competitive Scholarships and Fellowships?

Make sure to find good letter writers, because references are one of the most important parts of your application. If you are fortunate enough to have people at your school with experience reading over scholarship essays, their feedback on a polished draft will not only help your application, but will also improve your writing skills. It only takes one dissatisfied reviewer to knock you out of the competition, so be as conservative as possible with your essays. Finally, always keep in mind what the reviewers will be looking for, such as emphasis on broader impacts for the National Science Foundation Graduate Research Fellowship, hard-core research for the Hertz Fellowship, etc.

Max, Churchill Scholar, Goldwater Scholar

Conclusion

Chugging through college may be difficult, but don't let school distract you from preparing for life after graduation. Rather, make it work for you. College is the perfect time to prepare for your future profession. It's up to you to learn about potential career paths, network with others, prepare applications, and apply for fellowships. Your future will thank you for taking the initiative.

7 STEM in the Real World

In this chapter, we discuss a number of opportunities available to you with a background in STEM. Your STEM education will open many doors after graduation. Perhaps the sciences will grow on you during college, and you'll eventually realize you couldn't do non–STEM-related work the rest of your life. Or perhaps you'll fall in love with an entirely different field. Whatever the case, the lessons you will learn in your STEM major will prepare you for many fields both within and outside of the sciences.

Graduate School in the Sciences

A graduate school is an institution that awards advanced academic degrees. You might consider going to graduate school if you feel that further formal education will improve your job prospects or deepen your knowledge of a field you love. According to the US Bureau of Labor Statistics, students with graduate degrees earn significantly more money and experience less unemployment than those with only bachelor's degrees.[1] Graduate school in the arts and sciences is different than professional schools, which prepare college graduates for a specific career like medicine, law, or business. In comparison, graduate schools teach their students about more general fields of knowledge such as physics, sociology, literature, and so forth.

There are two main levels of advanced degrees that graduate schools offer: master's degrees and doctoral degrees. This section will focus on graduate schools in the sciences, but even if you decide to go to graduate school in a nonscience field, much of this information will still apply. To search for potential graduate programs:

- Solicit the suggestions of your current professors and graduate students at your university for reputable programs in your scientific area of interest.
- Read over recent papers featuring topics that you would like to study, and make a note of the principal investigator and his or her university or institution.
- Browse departmental and laboratory websites of the programs that interest you.
- Check out an online graduate program directory such as www .gradschools.com.

Applying to Graduate Programs

Unlike the college admissions process, graduate schools do not have a centralized application, which means that you need to look closely at individual programs to make sure you meet the requirements for admissions. Carefully read through the program websites and organize their admissions requirements into a spreadsheet to keep track of your application. We describe the most common components of the graduate school application below.

1) Letters of recommendation: Academia is a small world, and people studying similar subjects tend to know one another. You will want strong letters of recommendation from academics in the field who ideally know you well, have mentored you in research, have a good reputation in the field, and know the professors in the program to which you are applying. The program that you apply to will be much more likely to admit you if someone they know is vouching for you with specific examples of your potential.

2) Personal statement/admissions essay(s): These essays are an opportunity to differentiate your application from those of others, and you should aim to answer the following questions:

- Why are you interested in this field?
- What relevant research experiences do you have?
- Whom do you want to work with and why?
- Why you are interested in this particular program?
- What are your research goals, and what do you hope to do with your degree beyond graduate school?

3) Graduate Record Examination (GRE): The GRE is a standardized test required for admission to most graduate schools, similar to the SAT and ACT college-entrance exams. The GRE takes roughly four hours to complete and assesses ability to write, perform high school–level math, and understand written passages. There are also seven GRE subject tests offered three times per year that test knowledge of a particular field. Some

grad schools may require that you take one of these subject tests if it is relevant to that discipline. You can study for the GRE by purchasing prep books and taking practice tests or by enrolling in a GRE prep course. Some grad schools may have GRE cutoff scores for admissions. Scoring high on the GRE is great, especially for top programs, but graduate programs are generally more interested in knowing that you have the specific knowledge, experience, and drive to succeed in your field of interest.

4) College transcript: Good grades are one indicator of potential success in grad school. Grades in courses that relate to your discipline of interest are especially important. You don't need to have majored in a given field in order to apply to grad school in that field, but sometimes the courses you are required to take in a major may be less extensive than the grad school requires or expects. Plan early to make sure that you graduate with any and all prerequisites you need for your hoped-for graduate program. The average GPA to get into most doctoral programs is above a 3.5.[2]

The application process will begin about a year before matriculation. You should plan on taking the GRE in the summer one year before you intend to matriculate to graduate school, to complete applications during the fall, and to attend interviews (if your program offers them) in the winter. Elisabeth, a Dartmouth College graduate and a Fulbright Scholar, suggests that students interested in going to graduate school should get research experience as early as possible: "If you do so and you stick with the same research advisor/professor, build your relationship with them and get to know them. Try to get experience doing what you think you'll do after college in your internships or other extracurricular activities."

What Is a Master's Degree?

One graduate degree that you might consider pursuing is a master's degree. A master's degree signifies that a student has achieved advanced knowledge in a field, beyond that which is obtained in a college major. Master's programs will involve advanced coursework in a subject and—depending on the program—the completion of a master's thesis, a piece of well-thought-out original research. Most master's programs look for applicants who have done well in the college courses relevant to the field, obtained strong GRE scores, received good letters of recommendation, and conducted relevant research experiences.

Typical degree programs will last from one to three years. By the end, students will receive either a Master of Science (MS) or Master of Arts (MA) degree, depending on their program.

For some career choices, like architecture, this couple of years of study is necessary to demonstrate sufficient knowledge to do the job. For other careers, a master's may not be required. Find out which is the case for your particular career interests, what the graduates of the master's programs you are applying to typically go on to do after graduation, and whether you can transition into a doctoral degree program after finishing the master's degree.

Financial aid and scholarships can be difficult to come by for master's students. Seek out scholarships aggressively and fill out financial aid applications on time, but be aware that you may have to end up footing the bill for the program yourself.

What Is a Doctoral Degree?

Contrary to what the name "master's" might imply, it's really the doctoral degree (PhD) that indicates the highest level of mastery in a specific discipline. A person with a PhD is entitled to call himself or herself a doctor (not to be mistaken for a physician, which requires a different advanced degree). Students who are passionate about research in their field of study may consider obtaining a doctoral degree. A PhD is almost always required to be the head of a laboratory or research group and become a tenured faculty member at a university.

A PhD program typically lasts four to eight years and involves the completion of two years of coursework followed by several years of independent research. In the initial years of a doctoral program, students will take courses, rotate through labs they might consider working in, and prepare for examinations that demonstrate in-depth knowledge of the field. The successful completion of the last of these examinations, often called the *qualifying exam*, will allow them to begin their dissertation, the culmination of all their education in the form of an original research project. The student, now referred to as a doctoral candidate, will work on the dissertation under the guidance of a research advisor. When the project is complete, it will be presented to a committee of faculty members who will decide whether or not it is of adequate quality to allow the doctoral candidate to be awarded a PhD.

The Realities of a Doctoral Degree

The decision to enroll in a PhD program requires a serious commitment of time and money—both through years of lost earning potential and possible tuition or living expenses. The dissertation can stretch on and on until you give up, run out of money, or finally complete it. Ten years after enrolling in a PhD program in STEM fields, only about 60 percent of the students will have completed their degree while 30 percent of students will have dropped out, and 10 percent of students will still be attempting to finish their dissertation.[3] These depressing statistics are a reminder that you should consider getting a PhD only if you know what you are getting into. Think carefully about a PhD program before applying and potentially spending many years of your life working for uncertain reward. Many potential PhD students take one or more gap years after college to conduct full-time research or try other things before graduate school. This can be a time in which to think about whether or not to do a PhD as well as a time in which to strengthen your application. Have frank discussions with current PhD students and recent graduates from PhD programs about their experiences and their thoughts about the job/academic market now and in the future.

Some doctoral programs allow students to earn a master's degree a few years into the program after completing certain academic milestones. If you can get a master's degree, you at least will not leave graduate school empty-handed if you decide not to finish your dissertation.

Deciding on a Doctoral Degree

How do you know whether a PhD is right for you? The answer is "by doing research." This is highly important both for getting into a PhD program and for knowing if you want to be there to begin with. If you can get your research published that is ideal, but it is not necessarily required. The ability to work full-time on independent, self-directed, and self-motivated research is exactly what will be needed to finish a dissertation. The ambiguous and unstructured nature of a dissertation is unlike previous educational experiences. In college, your problem sets and tests all had clear answers. Real research offers no such assurances, and you should be comfortable with this ambiguity before deciding to devote a significant part of your life to it. Research requires you to cope with many, many failures—for each scientific experiment that works there are a lot of broken test tubes, dead lab rats, and ripped-up pages in a recycling bin. The faculty who are admitting you are interested in making sure you can handle research; after all, they are hoping that your research will help bring them lots of shiny publications that will help them get grants/tenure/prestige. In a graduate program, you will confront challenges and failures over and over again, and for this reason, it is important that you have the professional and the financial support to get yourself through the program.

Finding a Research Advisor

As a PhD student, you are bound to your advisor for the duration of your dissertation, and you need to choose him or her very wisely. You need to mesh at a professional if not a personal level. Sometimes you will be matched with your advisor upon entering the school and other times you will choose the advisor a few years into your program.

Even if your PhD program gives incoming students some time to choose an advisor, you should already have an idea about whom you might want to work with while applying to the program. Additionally, prospective PhDs should have a clear idea about what research topic to pursue before applying to graduate school. It is not enough to know that you want to study chemistry; you need to know, for instance, that you want to study methods for calculating the thermodynamic properties of polymers. When looking for advisors, ask yourself:

- Are there faculty members at this program who study the topic I am interested in?
- What are these faculty members like?
- What are their reputations in the field?
- What did their previous students think of them?
- Where are their previous students now, and how many years did it take them to get their degree?
- Did their students publish in any high-impact academic journals?
- Did the advisor overwork her students as a means to an end, or did she take some time to help them develop professionally?
- Did the students all graduate and move on to great jobs?

Contact potential advisors BEFORE applying to a school to see if they may have spots for graduate students in the future, and express your interest in working with them. Doing this will save you from wasting a lot of time and money applying to a place where you won't be able to do the work you want to do. If you have the opportunity to interview at the school, meet potential advisors and their current graduate students and be prepared to ask questions/speak intelligently about their research and your own research. In addition, for every institution you apply to, make sure that there are a few advisors whom you can see yourself working with. Don't put your eggs in one basket by going to a school that has only a single faculty member whom you would want as an advisor. If that person leaves the school, doesn't get tenure, or runs out of money, you would consequently find yourself in a very difficult situation.

Financial Support in Doctoral Programs
Many STEM PhD programs provide funding for tuition and a stipend for living expenses. However, funding for PhDs can

be uncertain. In the humanities, PhDs usually have a particularly hard time getting funding. Even before you apply to a PhD program, search funding databases and apply for grants, scholarships, fellowships, and other sources of funding for your future education. Institutions like the National Science Foundation (NSF) and National Institutes of Health (NIH) provide major sources of funding for PhD students. If you secure funding before you apply for a PhD program, you are automatically a more attractive candidate because you are funding yourself rather than making the school fund you. Even when you begin your PhD, you need to keep looking for funding opportunities both within and outside your department.

Most PhD programs offer teaching assistant positions where grad students are paid to teach or assist the professor in undergraduate classes in order to earn their keep during grad school. This earns money for the student, but can be a distraction from the work to finish their dissertation. Many grad students prefer to get research assistantships, which are paid positions to do research—often the research that they are already doing for their PhD. Find out what types of funding your department might be able to provide you with and whether or not this funding is guaranteed for the duration of your PhD.

Funding is also a factor that you should use to evaluate potential advisors. What sources is your potential advisor using to fund her lab/research group? Are those funds likely to run out soon? Is the funding already earmarked for a certain project? Will the professor be able to provide you with some funding or at the very least continue to fund her own research? Any given source of funding will only last for so many years, and if your PhD goes on longer, you could end up needing more. Can you live on the funding you will receive?

Some 30 to 40 percent of STEM PhD students end up taking out loans during their graduate education.[4] Ask how many students in the program end up going into debt. Many students have to live frugally during their PhD. Apply carefully to schools and scholarship programs that can take care of you financially. If you can't find a suitable financial arrangement, consider strengthening your application and waiting to apply in a later application cycle.

Don't be too discouraged by these warnings. Graduate school is absolutely the right choice for many students. However, this path is not to be traveled just because you don't know what comes after college.

Figure out how your PhD will benefit your future. A master's might be better than a PhD for certain careers because the length of training in a PhD program is too long and the type of training too specific for the work that you will do on the job. Not to mention, you are losing income by going to school when you could be working.

Many PhDs aspire to become professors, but simply getting a PhD guarantees nothing. Tenure-track positions in academia are extremely competitive. Most PhDs who wish to become professors will typically complete years of additional training as a postdoctoral fellow in order to get the publications that will make them competitive for a professorship.

A doctoral degree gives you expertise in your area of study and can pave the road for a number of career paths in the sciences. Regardless of your future intended career, the degree will give you credibility as someone who has studied deeply and attained the highest level of education in your field.

Careers in Medicine and Healthcare: Applying STEM to Care for Patients

Medicine is the application of science to the improvement of health. The road to becoming a healthcare professional can be long, arduous, and expensive, involving many years of schooling and training. All of this is intended to prepare a student for the heavy responsibility of caring for another person's life. The rewards of being a healthcare professional include job stability, good salaries, and the enjoyment of using science to take a direct role in helping others.

Many college students have an interest in being "premed"—that is, going on to medical school and becoming a physician. Consequently, we have included a large section on preparing for and getting admitted to med school. This requires a lot of planning and hard work; with so many interested students, there can be a lot of competition and "weed out" classes. For this reason, students interested in the health sciences should also be aware of some of the other medical professions that exist.

A brief side note: There are two types of physicians in the United States, Doctors of Medicine (MDs) and Doctors of Osteopathic Medicine (DOs). Both types of physicians are fully qualified to take care of patients, but the medical schools they attend give out different degrees. The differences between MDs and DOs are far outnumbered by their similarities, but you should be aware of the fact that MD medical schools are usually considered more prestigious and are tougher to get into than DO programs. Additionally, while the term "doctor" is usually used to refer to physicians, some other healthcare professionals have doctoral degrees in their own fields. For instance, pharmacists are doctors of pharmacy (PharmD) and dentists are doctors of dentistry (DDS or DMD).

Other Healthcare Professions

A number of other healthcare professions involve taking care of patients, making substantial salaries, and completing an education similar to that undertaken by physicians. Admission to their respective training programs is typically less competitive than admission to medical school. Such careers include dentistry, podiatry, pharmacy, physician assistant practice, and nursing.

For historical reasons, doctors who treat the teeth and the feet, dentists and podiatrists, respectively, go through a separate training process than doctors who deal with other parts of the body. If this does not make sense to you, then you are in good company.

Pharmacists are responsible for dispensing medicine to patients. This requires a solid understanding of chemistry and the human body, as well as a knowledge of how different medications might interact with the body and with each other.

Physician assistants (PAs) are highly trained medical professionals who do much of what a physician does, including examining patients, ordering tests, and diagnosing. "Physician assistant" isn't a very descriptive name for the job; PAs may practice under the supervision of an MD or DO, but they can also work independently, depending on state regulations.

Nurses are a very heterogeneous group of healthcare professionals. Many nurses train for their career with an associate degree (usually a two-year college degree) or a bachelor's degree specifically in nursing. However, nursing could still be an option for someone who has already graduated from college if he or she were interested in pursuing further schooling. Nurse practitioners (NPs) have the greatest scope of practice among all levels of nursing, and must complete graduate or doctoral-level training and board certification.

As should be clear to you by now, there are many different routes to caring for patients. Students with an interest in potentially becoming a dentist, physician assistant, podiatrist, or pharmacist should note that much of the scientific coursework required for medical school is also required in order to get into these professions. These careers will also require applications and years of training in advanced professional degree programs, passing scores on standardized national exams, and closely supervised on-the-job type training.

Those who are drawn to healthcare but, for whatever reason, prefer not to work directly with patients may also consider working in public health or healthcare administration. A Master of Public Health or Master of Health Administration are degrees often sought by professionals with interests in these fields.

Getting into Medical School

The application process for MD- and DO-granting medical schools is quite similar. Both types of schools use their own "Common Application" style websites (AMCAS and AACOMAS, respectively) through which all applications are submitted (note: some medical schools in Texas use a system called the TMDSAS). Your college's pre-health advisor, a number of books (e.g., *Med School Confidential*), and the website of the American Association of Medical Colleges (AAMC) can give you a detailed look at applying to and preparing for medical school. The AAMC's "Aspiring Docs" blog offers some great insight as well. We highly advise that you meet with your school's pre-health advisors early in your college career in order to establish a rapport with him or her and to familiarize yourself with the challenges that applicants from your college face when applying to medical school. What we would like to do here is give a brief overview of what is most important to know about this process.

In brief, a student is prepared to apply to medical school after satisfactorily completing the pre-medical coursework and taking a test that assesses mastery of this coursework, called the Medical College Admissions Test (MCAT). The exact pre-medical course requirements vary a bit from school to school, but they usually involve a year each of college level biology, general chemistry, organic chemistry, biochemistry, physics, mathematics, and sometimes English. The MCAT tests proficiency in all of these subjects as well as knowledge of sociology and psychology, so further coursework in these two fields may be useful for test preparation. Because of the scheduling involved with taking all these courses, planning out your coursework early is important especially if you want to graduate on time, major in a nonscience subject, or take on another major. There are no required majors for pre-medical students or preferences for particular majors, but roughly half of all medical school applicants choose to major in biology.[5]

Those who want to attend medical school but were unable to take the pre-medical courses in college or want to boost their grades should consider enrolling in a structured or unstructured postbaccalaureate program. Recent college grads and graduates from years back can enroll in these programs to learn all the science they need to start medical school in

1–2 years. A structured postbaccalaureate program features a defined set of courses and is primarily aimed at students who have not previously taken the prerequisites for medical school. You can also take an unstructured postbaccalaureate program by taking courses in your local 4-year university as a non-degree student. These programs can often be as pricey as undergraduate tuition, but it may be worth it for those with a sincere interest in a career as a physician.

The application to medical school consists of three parts:

- Primary application
- Secondary applications
- Interviews

Applicants to medical school will submit their primary applications about a year before they hope to matriculate. In the primary application, applicants need to fill out a lengthy form about their extracurricular activities, write a personal statement explaining who they are and why they are interested in medicine, submit a transcript and MCAT scores, and request letters of recommendation from professors and mentors. The sooner you submit your medical school applications the better—many schools use rolling admissions.

After the primary application, each medical school will issue a secondary application containing school-specific questions. Some schools will send secondary applications to every single applicant, whereas others will extend them to their top applicants. Finally, the most impressive candidates will be invited to the medical school campus for an interview. The interview process can consist of a traditional sit-down-and-tell-me-about-yourself interview or a series of multiple mini-interviews (MMI). In an MMI, applicants' ethics, empathy, critical thinking, and teamwork abilities are assessed by making them think quickly through multiple hypothetical questions and scenarios with different interviewers. For the MMI, the answers aren't as nearly as important as the steps you take to communicate your thoughts.

Some schools release their admissions decision a couple of weeks after the interview, whereas others release the decisions for everybody on the same day. Finally, if students are fortunate enough to have multiple acceptances, then they must choose the medical school that they will enroll in during the coming academic year by late April.

The average MD applicant ends up applying to 14 different schools.[6] Depending on how many schools you apply to, the combined cost of the MCAT, the primary and secondary applications, and the travel to interviews will typically add up to thousands of dollars. While AMCAS will provide financial aid to applicants who meet a specified income threshold, most applicants will pay the entirety of their application fees.

There are many highly competent students applying for the same spots. From 2011 to 2013 only 44.4% of med school applicants received an acceptance.[7] However, unlike in PhD programs, almost everyone who enters a US medical school completes his or her degree successfully.

Medical schools place an emphasis on a variety of different factors in order to choose whom to accept. The most important are grades (especially in science courses), MCAT scores, letters of recommendation, research experience, leadership and shadowing experience, community service, and demonstrated motivation for medicine. We will talk a little bit about these factors.

- Grades/MCAT: Good grades and MCAT scores are necessary but not sufficient to get into med school. In 2013, the average matriculating medical student had a GPA of 3.69.[8] Start preparing for the MCAT well in advance of your test date by purchasing an MCAT review book and taking practice tests. Some students will enroll in an MCAT test prep course—this option might be worth considering if you learn better in a group or benefit from more structured learning. Give yourself plenty of time to train for this marathon of a test.

- Clinical experience: How can you know you want to be a physician if you haven't had any experience in medicine? Answer: You can't. Spend some time shadowing healthcare professionals, volunteering in hospitals, helping out in nursing homes, and generally giving yourself exposure to the life that you are considering embarking on. This is as much to benefit you as it is to strengthen your medical school application. Make sure medicine is right for you. Only you can know this!

- Research: Research experience in any field, scientific or nonscientific, is looked upon very highly by medical school admissions committees. If you can get a publication or multiple publications, then that is icing on the cake.

- Extracurricular activities/volunteering/leadership: You get the idea. Do interesting things with your life. Help other people, demonstrate that there is something or a number of things that you care about and have devoted a significant amount of time to. It makes you an interesting human being in real life and in the eyes of the admissions committee that is reading your application. A big part of getting into medical school is showing you are a decent person who can be counted on to make a sincere effort to care for others.

MD/PhD: Double Trouble

Prospective medical students who wish to become researchers may consider getting both an MD and a PhD degree. Many US medical schools offer competitive joint MD/PhD degree programs, which typically last for 7–8 years and offer free tuition and even a stipend. While the free tuition is nice, it is unlikely to offset the salary you will lose by spending more time in school, so the decision to apply to one of these programs should not be purely a financial one. The MD/PhD program can definitely provide a secure footing for a career in research, but many MDs will also go on to careers in research as well.

Time Off before Med School

The average age of a first-year medical student is about 24. It is not out of the ordinary that it might take some time after college for students to get their résumés to a point at which they can get admitted to medical school. Increasingly, students will take a gap year (or years) in which they can beef up any weak area(s) of their résumés before applying.[9]

Students who do not get any acceptances in their first application cycle should contemplate what might have gone wrong, speak with their pre-health advisor, and, in some instances, get feedback about their application from the schools from which they were rejected. After figuring out how they can best improve their chances, these students may wish to try again in the next application cycle.

Life in Medical School and Beyond

In medical school, you will learn the fundamentals of human health and disease over the course of four years of intensive study. In the first 1–2 years, you will learn mostly in the classroom and laboratory, while in the remaining 2–3 years you will work in a hospital or clinic, getting on-the-ground experience with caring for patients.

Medical training goes on for a long time after medical school is finished. All students will study roughly the same information in medical school. In the last year of medical school, they will apply to a training program in the specialty of medicine that they will eventually practice (e.g., internal medicine, neurosurgery, pathology). This stage of medical education is called *residency*, and it consists of 3–7 years of caring for patients and developing essential skills under the supervision of fully qualified physicians. It is only after residency that you can become fully qualified to practice medicine independently. Some doctors will opt to complete an additional supervised training program of 1–3 years' duration, called a *fellowship program*, in order to subspecialize in a narrower area of medicine or surgery.

There are vast differences in the training and day-to-day life of physicians who choose different residencies. Consequently, if a career in medicine interests you, you should also get a sense of what particular specialties you might enjoy as well as how you can prepare for those specialties while in medical school. Some are much more competitive than others. In general, residencies will evaluate medical school applicants on the basis of their grades, scores on standardized exams, research experiences, the quality of medical school they attended, and other factors.

While most physicians practice medicine full-time, there are still many other options for those who complete medical school, such as careers in academic research, biotechnology, or healthcare consulting.

Careers in Teaching: Shaping the Next Generation

Picking up this book means that you're college-bound or in college, so you have already benefited from the effort that your teachers have dedicated to your education. If you want to mobilize your scientific education to inspire the next generation of students, consider teaching as a career.

There are a number of transitional, one- or two-year opportunities to explore a career in teaching that do not require additional degrees. Such programs include Teach for America (TFA), City Year, World Teach, AmeriCorps, Match Education, Citizen School, CoaHCORPS, Boston Teaching Residency, NYC Teaching Fellowships, Reach to Teach, and Inner-City Teaching Corps of Chicago.[10] These programs are often geared toward providing college graduates opportunities to serve in communities with relatively poor educational infrastructures. For instance, Teach for America's school sites are commonly in urban neighborhoods around the country, and Match Education operates at relatively new schools in the Greater Boston Area. As a first-time teacher with only a rudimentary background in education you will face a formidable challenge, but you will also be faced with the potential reward of connecting to and improving the lives of students from underprivileged backgrounds.

In order to develop a long-term career in teaching, formal certifications are required. Though each state has slightly different licensing requirements, the general licensing pipeline for teachers is about the same nationwide. For entry-level teachers, there is a temporary certificate (lasting a few years), requiring a bachelor's degree, course experience in pedagogy/ education, and demonstration of proficiency in one particular area of study. Following several years of experience, which may involve serving as a student teacher, one can pursue a more permanent license in teaching. Some states (e.g., California), differentiate between certification to teach in a single subject or in multiple subjects. Because each state is unique, check out Teach.org for the specifics of your location. Of note, there are differences in qualifications and training required for those teaching 7th–12th grade and those teaching K–6th grade. Teaching at the 7th–12th grade level will require greater subject material expertise.

Careers in Law

Traditionally, law school has not been a popular destination for undergraduates with hard science or technology backgrounds. In recent years, however, law schools have begun actively recruiting STEM majors to meet the increasingly complex legal needs of the nation's burgeoning technology and biomedical industry.[11] STEM majors enjoy some of the highest admission rates among law school applicants.[12]

STEM majors with an interest in the protection of intellectual property are in particularly high demand. For example, STEM majors are prized in the field of patent law, which concerns the rights of researchers and inventors to profit from their discoveries. Patent lawyers must often be able to interpret and compare highly complex design schematics in order to determine whether technology has been illegally copied—a task that STEM majors are uniquely well suited to perform.

Moreover, the explosive growth of the technology and biomedical industries, particularly in California's Silicon Valley, has created a niche for STEM-savvy lawyers who understand the technical details of marketable products and services and can anticipate potential legal problems arising from their development, distribution, and usage.

Just as graduate school applicants must complete the GRE, and medical school applicants the MCAT, law school applicants must sit for the LSAT, or Law School Admission Test. The LSAT tests reading comprehension, analytical reasoning, and logical reasoning through approximately 100 multiple choice questions, along with an unscored essay section.[13] LSAT scores range from 120 to 180, with scores above 172 usually representing the 99th percentile.[14] Unlike the GRE and MCAT, the LSAT is offered only four times per year. Although the exam can be taken multiple times, most law schools will consider the average of your scores over the past five years when reviewing your application.

Like the Common Application or the Coalition for Access, Affordability and Success used in undergraduate admissions, the law school application process is streamlined through LSAC, the Law School Admission Council. To apply to law schools, you will submit a personal statement, résumé, and (typically three) letters of recommendation to LSAC, which will then transmit those materials to your selected law schools. Some law schools—particularly the highly selective ones— will ask for additional materials to supplement your application.

Very few law schools will require applicants to undergo an interview. A notable exception is Harvard Law School, which offers a limited number of interviews to select applicants. Law school admissions are conducted on a rolling basis starting in the fall and closing at the end of April. Applying early in the process is highly recommended, because there will be fewer unfilled slots available in the spring.

The legal industry was hit hard by the recent recession. Many of the nation's largest legal consumers were forced to curtail their legal expenses, leading to mass layoffs at several of the nation's most prestigious law firms.[15] The resulting financial squeeze on young attorneys has led to a substantial drop nationwide in law school enrollment.[16] If you are considering applying to law school, be sure to make an honest and sober assessment of your prospects with your career counselor. While graduates of the nation's top law schools still enjoy relatively unencumbered access to prestigious and well-paying "Big Law" jobs, many graduates from middling schools struggle to find any legal work at all. Do not take on substantial debt to attend a law school unless there is a robust market for that school's graduates.

Following law school, you must sit for the Bar examination in the state where you hope to practice in order to become a full-fledged lawyer. Of note, there is a unique bar exam for those who wish to practice patent law.

Students Say: What Should STEM Students Consider about Applying to Law School?

Emphasize your research interests in your applications. Because there are relatively few STEM majors applying to law school, schools get particularly excited when they see a candidate interested in studying the intersection of scientific disciplines and the law. By playing up these interests, candidates might attract the attention of like-minded professors on the admissions committee.

Raza, Stanford Law School

Finding Employment
Science-Related Jobs

As a science major, you'll have a number of science-related jobs you can pursue after graduation. For those with BS or BA degrees, science jobs can be found primarily in three venues: academic institutions, government research facilities, and commercial corporations—the last of which is collectively referred to as "industry." We will discuss each of these job types briefly below. If you find yourself particularly drawn to a certain field, we encourage you to talk to your major advisors or look online for more information. Given the sheer number of different jobs in each of these fields, it is difficult to provide particularly specific information within a reasonable space.

For freshly minted graduates, employment at academic institutions is generally limited to research assistantships or lab technician work. The salary of these assistantships is usually quite small as they tend to be drawn directly from the principal investigator's grant budget. Most often, college graduates pursue assistantships as a means of boosting their résumé via authorships in peer-reviewed publications. With additional research experience, they can become more competitive candidates for PhD programs or medical schools. Academic assistantships are not long-term careers one can settle into. In the absence of advanced degrees, they rarely offer opportunities for professional advancement.

Employment at government research facilities provides better salaries and benefits than academic research work, generally. However, it is, likewise, not a long-term career path for someone with only undergraduate training. Just as you're generally required to have a PhD to attain funding and lead a research team in academia, you'll similarly find that growth opportunities at a government facility are greatly limited in the absence of graduate training. Like academic research assistantships, government research work for college graduates is a temporary work experience prior to pursuing additional education.

The third work option, industry, does provide opportunities for a long-term career. "Industry" encompasses a remarkably broad number of jobs. The term refers to essentially any industrial or private companies that require science knowledge either in manufacturing or research and development. In this sense, you could work for "industry" in the research and development office of Kellogg (developing new cereal products) or for Merck (creating new pharmaceuticals). Accordingly, the ranges of career experiences you can have in industry are exceptionally varied and difficult to concisely define. However, there are a number of factors that remain consistent in industry jobs. The most important is that you will have opportunities to advance your career in industry, even in the absence of an advanced degree. This works for two reasons. The first is that research has become more automated and accessible. PhDs are no longer required to run certain experiments that they may have been needed for in the past. For example, many genetic analysis techniques, which were once limited to PhD-trained experts, can today be run by automated machines operated by college students. In this way, opportunities for BS and BA level workers to lead research projects have become available in recent years. In fact, corporations are changing their business models to recruit and advance BS and BA graduates for this purpose.

Melissa Harper, vice president of global talent acquisition and diversity at Monsanto, a multinational agrochemical corporation, told sciencecareers.com that "[BS, BA, and MS scientists] are very critical—not just for our business, but for our industry. They fulfill needs of which a PhD is not necessarily required." Additionally, BS and BA graduates can find opportunities for advancement by taking on managerial and business responsibilities. Since BS and BA graduates can now lead research teams, the business side of operations becomes a viable career advancement path—you may one day be able to lead the company you were working for! In fact, many firms make a point of advertising the dedication to career development opportunities that their entry-level positions offer.[17]

There is a diversity of jobs within industry after graduation. Depending on your major and research experiences, very different opportunities become available to you. Some of the most common STEM industry jobs are found in biotechnology, energy, engineering, and technology sectors. For instance, you may have heard of such pharmaceutical giants as Pfizer[18] and Merck. Hires by such companies are usually focused on chemistry and biology majors.

Another large field within the scientific industry is the energy sector, with such companies as Shell and ExxonMobil. These companies, in addition to hiring chemistry and biology majors, hire engineering, geology, and physics majors. For example, one listing by ExxonMobil for an Exploration and Production Geologist and Geophysicist calls for a BS in geology or geophysics. The responsibilities consist of assessing potential new sites of drilling, helping maintain current production sites, and assisting in the development of new sites. Another listing for ExxonMobil asks for a Reservoir Engineer with a BS in engineering (chemical, mechanical, or petroleum engineering) to analyze and optimize the performance of oil or gas wells by studying well data and developing economic analysis reports.

Other fields in industry include engineering firms, such as Pratt & Whitney and Lockheed Martin, and biotechnology firms, such as Amgen and Gilead Sciences. Like energy and pharmaceutical companies, these firms require an application listing your science background in your daily work experience. However, as we hope is apparent, the types of industrial work in these firms can vary greatly depending on the role of the company and your particular job description. Some of the most sought-after STEM jobs are located within the technology sector. According to the 2015 *U.S. News & World Report's* ranking of the Best STEM Jobs, four of the top five were computer science–related—software developer, computer systems analyst, information security analyst, and web developer. While the merits and the utility of these lists are debatable, the high rankings reflect, in part, high salaries, low unemployment rates, and the recently growing demand for these careers. Given the progressive nature of technology, many of the job descriptions in this sector and the demand for these jobs will likely change rapidly over time.

As you begin to think about the kind of job you'd like to have after graduation, talk to the professors and academic advisors in your major department. Discussions with these individuals will help you figure out exactly what your area of interest or major qualifies you for in the current job market. They may be aware of a number of firms that have a record of hiring graduates from your institution, and will likely provide very specific direction in your job search. They may even know a few of the directors at such firms and may make some phone calls on your behalf. The job market is constantly changing, and it's difficult to say which kinds of companies will be hiring at any given time. However, it's important to figure out the general job fields that your major will qualify you for early on.

Nonscience Jobs

Many STEM graduates will pursue work in a nonscience field. In fact, the rigor of a science degree can give you an advantage when applying to nonscience jobs, if you have the relevant skills for the position, as employers will recognize the hard work required to complete a course of study in STEM.

There is a wide variety of fields that seek out science students. Given the sheer breadth of the job market and the limited scope of this book, we'd like to focus on two popular careers for successful STEM students: consulting and investment banking.

A management-consulting firm is a business that seeks to improve the performance of other organizations by analyzing their business practices and providing plans for improvement. As a consultant, your job essentially is to be a "logical thinker." You are asked to think logically through business problems. What is the best way for an oil company to expand its reach into a Frontier Alaskan town? How many seats should the stadium of a new NFL team contain? What should be the price for a new drug? Being a consultant can thus be an exciting line of work for science majors due to the creativity and challenge it offers.

In light of its problem-focused nature, management consulting is recognized as one of the best fields in which to begin one's career, even if you'd like to one day pursue something entirely different. You develop a strong business sense and more importantly the ability to think through and solve business problems. Initial offers for incoming analyst positions in consulting firms are approximately two years. Many analysts stay and become associates. Others go to business school and return to the consulting or finance industry. And there are others who start their own businesses or pursue other interests.

Investment banking is another high-powered field and an excellent if competitive job option. Investment banking is essentially the issuing of securities, or tradable financial assets for corporations and other financial entities, and then the management of those securities in order to raise capital for clients.

The salary for investment bankers is quite high. With your standard salary plus performance-based bonus, you could easily earn six-figures in your first year on the job as a banking analyst. In addition to the salary, an entry-level analyst position at an investment bank offers a strong place to begin a career in finance. Analyst contracts generally last for two years, after which many if not most individuals leave their firms either for business school or another financial field, such as hedge funds. A smaller percentage of analysts are retained by the firm and promoted to associates, a position that has greater responsibility and pay.

However, despite its prestige and high pay, investment banking is notorious for being an exceptionally intense career. It's entirely common for analysts to burn out prior to the expiration of their two-year contract and seek out other job opportunities. As most analysts will agree, for all the money they earn, they rarely have time to spend it.

Entrepreneurship: Bringing Your Idea to Market

Your career path is not just limited to getting additional schooling or joining an existing business after graduating from college. After all, Facebook and Microsoft were created while Mark Zuckerberg and Bill Gates were in college. Numerous other familiar companies like *Time* magazine, Dell, reddit.com, and FedEx were conceived of even before their founders had received their bachelor's degrees.

Entrepreneurship comes in a variety of packages, and talented college students all over the country have started businesses ranging from gaming apps to ergonomic serving trays. At its core entrepreneurship is the process of starting a business, gathering resources, and taking on the responsibility of risks and rewards. As a science major, you are equipped with some of the hard skills needed to address the needs of the market.

Why You Should Be an Entrepreneur during College
1) College Is the Most Flexible Time in Your Life

While college demands time and attention, you have greater flexibility in your schedule than people working nine-to-five jobs. If you factor in winter and summer vacations, you have lots of free time on your hands with minimal responsibilities. Plus, without ties to mortgages and dependents, you can live on the cheap and devote your free time to your business.

Delian—the cofounder of Nightingale, a Thiel Fellow, and an alum of Y Combinator and StartX—encourages STEM students who are interested in entrepreneurship to start now: "Play around with things even if they are side projects during school. Understand what sort of ideas excite you and start thinking about where to apply your skills and what you are learning to create a business you can influence in the world. School is awesome, but if you're not changing the world somehow with it, then what's the point?"

2) You Have Incredible Access to Professionals and Resources

Building your business during your college years means that you have the knowledge of your professors, access to workshops, and an alumni network for potential support. Depending on your institution, you may also be able to take business and technical courses and join entrepreneurial organizations that provide an ecosystem of positive peer pressure right at your college. Furthermore, creating a STEM-related venture allows you to apply what you've learned in class to real-world enterprise, gaining business skills in the process that will be invaluable long after graduation.

3) You Can Fail

One of the most compelling reasons for starting now is that you have less to lose if you fail at a business venture while you are still early in your career. This is not to trivialize the difficulty, stress, and financial burden that can be involved with failing at an entrepreneurial venture, but only to emphasize that it may be easier on you to fail early and get a new experience than it is to fail later on when your life is more established. In addition, assuming that you have kept up your academics, you will be graduating with your bachelor's degree as a solid backup.

Caveat: Why You Shouldn't Be an Entrepreneur during College

By embarking on a startup, you will be splitting your time between two sizeable responsibilities and competing with people who devote all of their time and effort to a single task. By struggling to balance both your academics and extracurricular activities, you may find yourself unable to take advantage of other opportunities that college has to offer. Furthermore, you may risk your chances for jobs and professional programs that place a strong emphasis on one's GPA, such as top financial firms, medical schools, and law schools. One simple answer to this problem would be to hold off on your project.

However, if you don't think that you can wait, you may be able to work with your school administrators to lighten your schedule or to take some time off from school. The next section describes the different paths that three young entrepreneurs took in order to pursue their business passions.

Case Study #1: Alison, Spiral-E Solutions, LLC

Classes were Alison's top priority. Or at least, they used to be, before she started her company.

During her sophomore year, Alison took an introductory engineering course in which she would join a team to define and solve a problem by designing a system or device. After brainstorming various ideas, students would draft a business plan, just as they would do if they were taking their product to market. In the end, the students would build and test a prototype and present it to a professional review board.

The theme for the term, the professor announced, would be bioengineering.

After a week of brainstorming, Alison's teammates were stymied. "They were saying, 'We can't find a project, let's just reinvent the crutch or the wheelchair.' That sounded boring. I didn't want to spend nine weeks of my life doing that."

Alison hopped on a bus to the nearest hospital and started approaching strangers in white coats to ask about the most pressing problems in their fields. Soon, she found herself sitting with the chief of the gastroenterology department.

The doctor noted that the biggest problem he faced in his practice was stabilizing the stomach during endoscopic surgery to prevent movement that could lead to accidental bowel perforations, but biotech companies had already invested millions in finding a solution to this problem. He pointed out this may not be a suitable problem for a small group in an introductory engineering course.

Alison was undeterred. She pitched the stomach stabilization idea to her teammates, who agreed to adopt the plan. Her professor, however, was unimpressed. Alison and her team received a "D+" on their proposal.

"Our grades were in jeopardy, but I told my teammates that this was fine and that we'd solve that problem, negate this grade, impress the class, and get an A for the course."

The group figured that vacuum suction, which was already being used in heart and cataract surgeries, could stabilize stomach tissue. They built prototypes with tubing and a shop vacuum, and transported their prototype to the university-affiliated hospital, where they successfully lifted and stabilized a whole pig stomach. Follow-up tests confirmed that the suction did not kill the cells. At the end of the quarter, Alison and her teammates gave the pitch with their supporting evidence, received an "A" for the course, and won the award for the most patentable and marketable device in the class.

Buoyed by her success, Alison met with Gregg, the director of the university's entrepreneurial network. He taught Alison the basics of starting a business, suggesting that the group sign an operating agreement so that there would be no issues of equity, file the necessary paperwork to become a limited liability corporation, file taxes, and apply for a provisional patent. The provisional patent allowed a breathing space of one year to file a real patent, so that Alison and her team, now called Spiral-E Solutions, LLC, would have time to raise the money for patenting fees and legal representation, altogether about $10,000.

Alison again turned to Gregg, who gave her a list of contacts to pitch her product to in the search for an investor. However, for the next half year, things did not go as hoped for Alison.

"Investors thought that this was either a terrible idea or pointed out the fact that I was inexperienced—which was true. At that point, I didn't have a financial background or a business background. I didn't have a team or a board of directors. This company was a one-woman show and no one took me seriously, until Scott came along."

Scott was an acquaintance of Gregg who created and sold medical device companies for a living. Alison and Scott met at 6 a.m. at a local diner.

"It was hard to talk about stomach stabilization during breakfast, but he understood it intuitively, so we started talking about engineering and my future plans. I told him that I wanted to be an entrepreneur, and he says, 'Great, I'll mentor you.'" At that point, that was more than anyone else had offered to Alison, so she began sending Scott drafts of business proposals for edits.

Alison then caught wind of a business plan competition hosted by the business school at her university. She won second place and $5,000 in the undergraduate division, giving her a shot at the main business school competition, which she presented in front of a group of three hundred people.

"I was sweating bullets. Everyone else had come in teams, but for me, this was a one-woman show, and I was giving a presentation in front of judges who were very wealthy because they had succeeded in businesses. In the end, I gave my pitch, and I felt pretty good about it."

The judges seemed to agree. Alison placed first and won the $25,000 grand prize, more than enough money to file for a patent and to convince Scott to come on board officially as a business partner.

Alison was featured in the *Boston Globe*, and investors finally began paying attention to the company. With their new equity, the team was able to develop a better prototype, and at the time of the interview, Alison had continued her venture by establishing a website, hiring an accountant, and forming a legal team for her company.

The project is at "Series A" fundraising, which is the first round of fundraising before getting the product tested by the US Food and Drug Administration.

"Winning $30,000 is great, but for the FDA, we're talking about a magnitude more of money."

While looking for investors for her company, Alison has also embarked on other projects.

Alison admits that after starting her company, her classes became less important because she saw them solely as tools to do something for the sake of her company. After talking with administrators, she received special permission from her dean to take a lighter course load.

"I graduated later than I would have, but I have no regrets. There's no way that I could do the company otherwise.

"Engineering was the only major on campus that I could get credit to build things. I was able to embrace that so I could become who I was supposed to be. I figured it out as I went and being comfortable with that was awesome."

Case Study #2: Christopher, Thiel Fellow

Despite the freedom and the vast resources that a school can offer undergraduate innovators, some students have found college to be too restrictive and time-consuming and decided to quit school and to gather their resources elsewhere.

Roughly a month into his first term as a student, Christopher knew that he wouldn't be in college for much longer.

"I found the curriculum that was attached to my college to be pretty restrictive in terms of what I wanted to do, which was to make mathematical artwork that was too mathematical for the art department and too artsy for the math department."

He talked to his parents, who agreed to let him take a year off during which they would provide him with financial support, but under the condition that he would finish his freshman year. He quit his classes, but he stayed on campus to maintain access to the school woodshop and the metal shop in order to develop his art portfolio. Before his year had ended, Christopher met a couple of students who were searching for a lead developer with serious design experience for their gaming company, Puddleworks.

"I jumped on the opportunity because it's not every day

where you come across people who are looking for someone with your exact interests and skill set. As the cofounder of Puddleworks, I ended up with equity and I had a real stake in what we were building and I designed our first products."

Christopher began working on a physics-based puzzle game in the style of M. C. Escher called "System of the Mind," which uses gravity to navigate a complex spatial environment that incorporates mathematical concepts such as spatial paradoxes, non-Euclidean spaces, recursions, and strange loops. "The project scaled up because it was a lot bigger than I originally planned, which is partly my fault because I ended up making something that was personally artistic."

Meanwhile, Christopher was quickly advancing through the stages of consideration for the Thiel Fellowship, an entrepreneurial award created by Peter Thiel, the angel investor behind Facebook and the creator of PayPal. The fellowship would provide a no-strings-attached grant of $100,000 for two years and access to Thiel's vast professional network for applicants under twenty to skip college and focus on work, research, and self-education. By April, the application pool had been cut down to forty individuals, out of which twenty students would be selected as the annual Thiel Fellows.

"I thought that there could be some astronomically small chance that I could win this, and once I started getting further in the rounds, it seemed more like a reality. By April, I was pretty sure that there was a shot that I was going to get this. And I did." By becoming a Thiel Fellow, Christopher was finally free to work on whatever he wanted without worrying about finances.

With financial support for two years allowing him to work on whatever he wants to do, Christopher began working on a new product to visualize the harmonic relationships within music. He was inspired by his interest in the isomorphic keyboard, a musical input device consisting of a two-dimensional grid—rather than the linear array of a piano—of keys arranged in a way in which sequences of similar musical intervals maintain the same shape.

"When I was younger I tried to learn how to play the guitar and the piano several times, but I got so frustrated every time because it seemed arbitrary to me. Music is such a great system that seems to be hidden behind all of this music theory and muscle memory. If you are playing different scales on a piano, you are changing keys, which means that you are playing one of twelve different shapes for each one of those keys. Because there are the same shapes for different scales on an isomorphic keyboard, it is easier to play, and you can see the geometry becomes visceral because you are playing on a two-dimensional scale rather than one-dimensional."

Christopher began working on a program to visualize the harmonic relationships in any music and combined it with an isomorphic keyboard layout rhythm game for the tablet to develop an app to teach people how to play the instrument.

And while Christopher thinks that both of these projects will keep him occupied for his two years as a Thiel fellow, he is pretty certain about not returning to school.

"There's a lot of things that college gives you that are made easier. You can get any class materials; you can get access to smart people who can teach you things. Of course, you can get access to all of these things outside, but it's easy if you have an infrastructure that's there to provide it for you and to guide you.

"But in some ways, I shot myself in the foot for going into

my college because it offered such a rigorous and traditional curriculum. It was serendipitous the way that it worked out. Knowing what I know now, I would do the same thing, but looking back, I definitely took some chances that worked out."

Case Study #3: Matthew, Refresh Innovations Inc.

Matthew was no stranger to entrepreneurship. He had co-founded his first company, a photography and videography studio, at the tender age of 12. He scaled-up and innovated his business throughout high school by becoming the first studio in his local area to deliver photo content to customers digitally through email. In college, Matthew double-majored in Electrical and Computer Engineering and Economics, with a background in computer-aided mechanical design.

He first connected with his eventual business partners— Collin, an upperclassman, and Anish, a mathematics graduate from Cambridge University—in his junior year through the university entrepreneurship club. While Matthew, Collin, and Anish all had prior experiences in creating or being part of startups, none of them had produced a physical product before. Ultimately, they settled on an idea that Collin had devised in Stanford Business School—a contact lens case that would fit in a wallet, which the team dubbed "Contact Lens Refresh Card." After identifying their product, Anish, Collin, and Matthew began conducting market research in earnest— interviewing potential customers, sketching out the business plan, and developing a working model of the contact lens case.

According to Matthew, "We were trying to make a product that was feasible. There's a big difference designing prototypes in class and manufacturing a product in the real world. In our case, we had to consider user safety and various regulatory practices."

Matthew partly credits his economics and engineering background for preparing him to work on his startup. "Through the economics major, I gained a degree of financial literacy, which was useful in developing and de-risking our business, and through the engineering major, I gained a basic understanding of knowing what to learn to solve a problem."

One way that Refresh Innovations Inc. approached their product was through the scientific method. They first created a hypothesis and then ran an "experiment" in order to confirm or debunk it.

"Basically, our hypothesis was that people were unsatisfied with the way they use contacts in everyday life. How would you begin to answer this question? For one, you can conduct a survey, which is a great way to find out whether there is market potential in an idea.

"We found that 86 percent of contact wearers found themselves in situations where they want to remove their contacts but did not have immediate access to a carrying case. That was a valuable number for our business plan and potential investors, considering that there are 38 million people who use contacts, meaning 32.6 million potential customers."

After spending money out of their pocket to launch the company, Refresh Innovations Inc., the team took their idea to the Duke Start-Up Challenge, a yearlong entrepreneurial competition. After receiving helpful feedback from judges early in the program, Matthew refined their business plan and went on to win the competition, besting more than a hundred other proposals and netting the $50,000 grand prize.

To aspiring college entrepreneurs, Matthew had this to offer: "You can't be afraid to fail, and you can't be afraid to ask basic—and sometimes radical—questions. Plus, even if you fail, you'll never end up worse off than before."

Tips for College Entrepreneurs in Creating and Running a Startup

1) Construct a Business Plan

A business plan is a document that charts the future trajectory of your startup. While it need not be formal, take care to outline some of these basic components:

- Mission statement: What are you trying to accomplish?
- Description of your product or service: What product/service will your company offer?
- Market analysis: Who are your competitors, where do you fit in that niche, how are you different from others, who is your target demographic, and how is the market expected to grow in the future?
- Marketing: How will you advertise your product/service?
- Spending budget/financial requirements: How much money do you need to initiate and grow the business? How will you secure these funds?
- Goals and projected achievements: What are the milestones that you need to reach in order to sustain and grow the business?[19]

2) Prioritize Your Time

According to Alison, the first student in our entrepreneurship case studies, the most difficult thing about her time in college was running a business while being a student.

"There were days when investors would call when I'm in class or during midterms or finals. Balance is hard to do, and that would be true whether I was running a company or not." She could spend time with her friends on Friday for only a short time because she had to prepare materials for Saturday Skype conference calls. On some days, she'd be so wrapped up in her work that she'd forget to eat or sleep.

"The startup world is so fast-paced you can work all the time and no one is going to tell you no. The best lesson that I've learned in the last five years is that you have twenty-four hours a day, and if you don't allocate your time, something else will."

For Alison, dinner became her "sanctuary time." She didn't check her email or look at her phone; instead, she relaxed by watching *Family Guy*. "Even just answering emails for the company could be time consuming, so you have to create space, or you're going to work all the time."

3) Know Your University Regulations

While your institution may offer a plethora of resources to grow your business, Matthew, for one, took care not to touch some of the offerings at his university. "Check with your college's technology transfer office about their regulations regarding intellectual property or patent policy. There may be rules about what you can and can't do if you want to keep ownership over your property without getting the school involved. For instance, if you consult a professor and he or she gives you advice that leads to a new product, you may not be able to add their advice to your patent without the university taking a part in it."

Schools may have additional restrictions about starting a business through your campus or dorm room address. Be aware of what you are allowed to do on your campus so that you don't get blindsided later on.

4) Raise Capital

Raising capital can be one of the most difficult parts of starting an entrepreneurial venture. Even before you can start asking people to part with their money to support your venture, you will probably need some initial funds to make your startup more interesting for investors or to get a loan.

One fundraising opportunity for college students is participating in a business plan competition. Refresh Innovations Inc., for instance, won the Duke Start-Up Challenge, netting Matthew and his partners a cool $50,000. Alison's Spiral-E Solutions, LLC, likewise, took in $30,000 from the undergraduate and the business school competitions at Dartmouth. Even if you don't win any prize money, a business plan competition will provide you with the opportunity to get feedback from the judges, many of whom are seasoned entrepreneurs and investors.

You may also be able to raise funds from incubators, organizations that aim to speed the growth and success of early-stage companies by offering legal support, office space, access to financing and professional networks, and mentorship. The business models of incubators vary; some incubators offer their services to entrepreneurs in exchange for equity—shares issued by a company—from the resulting venture, whereas others may simply charge a fee. Accelerators are a subset of incubators focused on growing businesses that are more developed, usually helping entrepreneurs to fine-tune the operational and strategic details of their businesses.

Finally, depending on the mass appeal of your project, you

may be able to bypass traditional means of fundraising altogether through crowd-sourcing platforms like Indiegogo or Kickstarter.

Careers in Business: Getting an MBA

A master's in business administration (MBA) degree can benefit individuals interested in following a wide variety of career paths. Most MBA programs take two years, with the exception of dual-degree graduate programs such as the MD/MBA path for healthcare professionals, or part-time/online programs geared toward increased flexibility.

Unlike law school and medical school, it's more common to gain several years of real work experience prior to entering a graduate business program rather than starting shortly after college. For example, the 937 students in the Harvard Business School Class of 2016 had an average age of 27 at matriculation.[20] So, if you are still early in your college career, business school may seem quite far off, but there are benefits to having this potential trajectory in mind early in your academic plans.

The first big question you might be thinking is, "Why should I put my career on hold to go back to school?" Spending additional years of classroom time may not seem appealing after having completed college and beginning your professional life, but an MBA could offer several tangible benefits:

- Increased earnings and job opportunities
- Potential for tuition coverage by your employer
- Ability and skills to start your own business
- Gaining professional leadership
- Networking/mentorship opportunities

As with most graduate programs, admission to business school is competitive. Many applicants will have real work experience and significant professional accomplishments.

Applicants to business school will have completed a degree at a four-year college or university and taken a standardized exam. For MBA programs, students are required to take either the Graduate Management Admission Test (GMAT) or Graduate Record Examination (GRE). The GMAT is a much more quantitative and problem-solving-oriented exam than the GRE, and the GRE is similar to the SAT in content and question style.

Because of the versatility of the MBA and the range of people who apply for the degree, many graduation paces exist to help you fit your degree into your future professional schedule. These include the following:

1) Accelerated: For those entering business school with ample business knowledge or for those who are in a bit more of a rush, over 90% of schools offer accelerated programs that can be completed in 10–15 months.[21] The median age for students entering these programs is roughly 29 (as of 2009), a bit higher than that of students in typical two-year programs.

2) Deferred Matriculation: Harvard Business School and the Stanford Graduate School of Business School offer deferred matriculation programs for college students. Applicants apply early in their senior year of college, and after being admitted, students are expected to gain at least two years of work experience before matriculating into the program. We mention these programs because they cater to non–business majors. Harvard specifically encourages applicants from science, technology, engineering, and mathematics backgrounds.[22]

3) 4+1 BA/MBA: During your senior year of college, your institution may allow you to begin your MBA coursework, and complete this coursework in a single postbaccalaureate year, saving you a full year of graduate school tuition.

4) Evening and Weekend MBA: If you would like to balance the pursuit of your MBA with continuous immersion in the workplace, these programs allow you to take classes afterhours.

Conclusion

A STEM degree can open doors to a wide range of careers. While certain degrees are more applicable to specific jobs than others, virtually all STEM majors will provide you with strong analytical and logic skills. These skills are invaluable in the real world and can be applied in many different careers.

8 In Conclusion

In this book we have written about the things that we hope will help you to reach your goals in college: making the most of your time (chapter 2), excelling in your STEM courses (chapter 3), choosing a major that's right for you (chapter 4), having a productive research experience (chapter 5), and preparing yourself for life beyond graduation (chapters 6 and 7).

But let's be frank, the most cherished memories of your college experience won't be of you acing that one test. Rather, it will be the spontaneous late-night conversation with your dorm-mates, that unforgettable night out in the city, or the course that completely changed how you thought about the world. There's a reason why you keep on hearing that these years will be some of the best times of your life. Depending on what you make of it, college can be a path to self-discovery, an opportunity to make lifelong friends, and the diving board to a new adventure.

So before we reach the end of the book, we propose six suggestions to help you get the most out of college.

1) Take STEM Courses Outside of Your Discipline

In 2003, Dr. Paul Lauterbur won the Nobel Prize for his part in the development of magnetic resonance imaging (MRI), now most commonly known as a medical tool used by radiologists to examine the inside of the human body. Considering the nature of the discovery, you might think he was a biologist or a physician. In actuality, he was a chemist with a strong belief in interdisciplinary thinking who used what had previously been thought of as a chemical technique to develop a tool that could answer medical questions. To kick off his Nobel acceptance speech, Lauterbur poked fun at the apparent strangeness that he, as a chemist, would be sharing a Prize in Physiology or Medicine. He went on to suggest that the convention of splitting up science into different fields is based on administrative convenience and that the different disciplines of science were not "natural categories with rigid boundaries to be defended against intrusions."[1]

The boundaries between scientific disciplines are fluid. In order to get a richer understanding of your field and of science in general, take a page from Dr. Lauterbur's book and enroll in a couple of mid- and upper-level science courses from other STEM majors. Use these as an opportunity to start combining seemingly disparate concepts and developing new problem-solving tools.[2]

2) Expose Yourself to the Humanities and the Social Sciences

College is an opportunity to expand your horizons and learn more about whatever it is that you want to learn about. If you can, take a moment to branch out and try different subjects in the social sciences and humanities. This may be the last time you have so many different classes you can choose from, so take a moment to enjoy yourself and learn something new.

In the words of Sydney, a neuroscience major from the University of Michigan: "Whether you go on to grad school or to work, your focus will be much more narrow after your undergraduate education. Don't stick with things you're not passionate about and don't hesitate to try something that interests you while it's available to you."

3) Learn How to Communicate Ideas

There's a saying at the Massachusetts Institutes of Technology: "Engineers who can't write work for engineers who can." You might not think that success in STEM has much to do with good communications skills, but, in fact, these skills are a necessity. Scientists and engineers must be able to explain complex topics to their colleagues, students, and the public and they must be able to convince others of the significance of their findings. If not shared effectively, even the most brilliant findings will languish.

For example, Ignaz Semmelweis was a Hungarian physician from the nineteenth century who observed that frequent hand washing could control the spread of disease.[3] This assertion seems obvious now; however, Semmelweis could never convince the contemporary medical establishment that his ideas should be taken seriously, partially because of the fact that he antagonized his colleagues and did not carefully explain his ideas. It took many years and the work of other scientists who were better able to explain their work in order to convince the world of the correctness of the germ theory of disease and the value of hand washing. Being able to persuade is sometimes just as important as being right.

Consider cultivating your communication skills, either by taking writing courses or through extracurricular opportunities. Many universities sponsor undergraduate science journals to help students develop the skills they need to communicate complex ideas into a concise article that can describe the importance of scientific findings to a wider audience. A science publication will provide an opportunity for you to reinforce what you have learned in class and give you a supportive, low-risk environment to improve your writing abilities by working under an experienced peer editor.[4]

If your institution doesn't have an undergraduate science journal, think about starting one by applying for student group funding or write about science topics for your campus newspaper. Also look into research blogs and undergraduate journals that take contributions from college students throughout the country, such as the *Journal of Young Investigators*.

4) Attend Science Talks and Seminars

Your introductory and intermediate STEM courses cover topics and ideas that have been widely accepted by the academic community. Science talks and seminars, on the other hand, will expose you to new research being done at the frontiers of your field.

In seminars, you'll typically observe scientists who are explaining years, if not a lifetime, of work. These new findings will be dismissed or embraced by their colleagues depending on whether they can be reproduced and whether they make sense within the context of future experiments. By attending science seminars, you can start identifying labs you might want to join or research topics that you would like to learn more about. If you come across an especially compelling presentation, go talk to the researcher after the seminar. Introduce yourself and explain your interest in the field. More often than not, you'll be one of only a handful of undergraduates at a science seminar and probably the only one who had the guts to approach the speaker.

5) Work on a Long-Term Side Project

In *How to Win at College*, Cal Newport—productivity guru and computer science professor—writes that students should "constantly be working on a 'Grand Project'" outside of their courses.[5] As we've pointed out in chapter 7, college is one of the best places to undertake new enterprises because of the abundance of time and the resources students typically have at their disposal. This project might be turning your research into a publication, turning an assignment in an entrepreneurship class into a business, or preparing the perfect application to law school. Try to make incremental progress on your big goal as you keep working to finish your degree.

6) Take Setbacks in Stride

In college, you'll rub shoulders with the smartest group of students you've met so far. You will compete with these same students for grants, jobs, internships, and leadership positions. As a college student, you will never lack for challenges—and, by extension, failures.

That's perfectly fine.

As you push against your boundaries, your limitations will come to light. In the best-case scenario, setbacks will reveal your weaknesses and how you determine your sense of self-worth. If you are honest with yourself and treat a setback as a springboard for growth, then you will become better equipped to tackle future challenges that you will face during and beyond college.[6]

In the words of one student whom we interviewed, Chris, a geology major at Lafayette College, when facing a setback "it is important to identify why you are disappointed, appraise whether or not it is realistic to be disappointed, and then make an action plan." If you realize that you are being disappointed about something unrealistic (e.g., about not having the highest GPA in your class) then you may need to modify your goal. However, if you think your goal is still possible, use the setback as an opportunity to take stock of where you are and make a new plan.

Students Say: Do You Have Any Last Words for STEM Students?

Go to campus events! Some of the things you will find most fun, you probably don't even know exist.

Alvin, Harvard University

Being involved was the most important thing to me in college and that got me a lot further than a great GPA might have. College is supposed to be four of some of the best years of your life, and if you don't keep that in mind, you have missed out! Work really hard and efficiently but don't forget to be involved and have a social life!

Robert, Northwestern University

Get started early. If you know you're interested in research, start contacting professors in your classes and in the department asking them about their research and ways that you could get involved. Be proactive. Professors are very busy people. Showing them that you're motivated and willing to work hard will take you a long way.

Daniel, Stanford University

Be social, and don't be afraid to meet new people in your class. It will make tough assignments a lot more fun and more often than not, you will get the assignment done faster when working together. Also talk to upper classmen about particular classes to see what sort of plans they have for after college. It is helpful to know where students can go with what degrees as you begin to think about the type of career you may pursue, and hearing honest opinions from people who were just in your shoes is extremely helpful in all facets of the undergraduate experience.

Colin, Dartmouth College

Closing Remark

As you go through your college education, continue to use this book as a reference when thinking about your major, your research, and your future career. Now that you've been introduced to some of the many obstacles you will face during college, we know that you will be able to overcome them with the knowledge you have at your disposal.

Appendix

Advice for Underrepresented Students in STEM

African Americans, Hispanics, and Native Americans constitute just over a quarter of the US population; however, they make up only a tenth of the science, technology, engineering, and mathematics workforce.[1]

Women, making up half of the population and half of all STEM bachelor's degree earners, are also underrepresented in STEM, comprising just 28 percent of the science, technology, engineering, and mathematics workforce in 2010.[2] Simply put, the demographics of STEM students and of scientists do not match those of the American population.

Why is there such a huge gap? In this chapter, we will talk about the most common obstacles faced by female, minority, first-generation, and economically disadvantaged college students and give tips and strategies to minimize these obstacles.

This chapter is written for underrepresented students and non-underrepresented students alike. If you are a reader who identifies with one or more of the groups we've mentioned some of the data about underrepresented students may sound discouraging, but simply informing yourself about the difficulties you might face will help prepare you for them. In recent decades, science students and professionals have become more diverse and there is every reason to believe that this trend will continue. A discouraging statistic does not take into account your drive, determination, and ability to succeed as an individual.

If you do not self-identify with any underrepresented demographic, read this chapter to better understand an ongoing problem in the scientific world and to better understand the perspectives of some of your classmates.

Challenges Faced by Underrepresented STEM Students

Subtle perceptions can have powerful consequences. A study in the 1980s found that students have different perspectives of the same essay if the author's name has been changed. College students who were randomly assigned to read essays written by John T. McKay, Joan T. McKay, and J. T. McKay rated the article more highly if it was attributed to John T. McKay, the fictitious masculine author, rather than Joan T. McKay. Interestingly, John was perceived as a superior author to Joan even when they were both writing about a topic considered to be "feminine," namely, the psychology of women.[3] In some cases, our work may be judged not solely on its own merit, but also through the perception that others have of us.

Some underrepresented students may even be affected by these biases themselves through a phenomenon called *stereotype threat*. When students find themselves in situations in which they feel they are at risk of confirming a negative stereotype about their own social group, they can become anxious about how they do, which hinders them from achieving peak performance on an assignment or a test.[4] Thus, negative stereotypes can seriously obstruct a student's academic performance; for instance, a female student who hears "girls aren't good at math" before walking into the testing hall may find herself feeling nervous about not doing well on a math final and potentially proving her bully right, despite being well prepared. Because of this additional source of anxiety, she may trip over questions and make mistakes she never would have made in the first place.

Challenges Faced by Female STEM Students

In 2005, Lawrence Summers, then-president of Harvard University, mentioned that he believed women lacked "innate abilities" in math, provoking public outrage.[5] Five years later, Dr. Janet Hyde, from the University of Wisconsin-Madison, led a study that looked through the test scores of seven million children.[6] Researchers compared the average scores of male and female students in various grades and the average scores of the students with the highest math skills. Dr. Hyde's team concluded that girls did just as well as boys and debunked Summers' hypothesis. Despite this, there are fewer female workers than male workers in a number of STEM professions.

If boys and girls have similar test results, then why are there more men than women in math and science? One contributing factor is unfortunately gender discrimination. A study in 1992 found that test scores of children at the end of the school year correlated with the teacher's expectation of student's abilities at the beginning of the year, regardless of whether the teacher's original expectation was true.[7]

These findings suggested that the misconception of science as a "masculine" field could unwittingly set teachers against female STEM students early in their education, and possibly later on as well.[8]

Female college students often report lower levels of confidence in their math and science abilities than male undergraduates, which leads to a decreased number of women who become STEM professionals.[9] Fortunately, gender gaps are closing; women now earn 48 percent of the undergraduate degrees in mathematics. However, in fields such as physics and engineering, the gap remains. These disparities persist when graduates begin their careers. Family planning, particularly in career paths that involving long stretches of training, can be particularly difficult. According to Dr. Donna Dean, the former president of the Association for Women in Science, for many female scientists, "there is never an obviously convenient time to have children. Each person has to sort out the timing that will work best for her overall."

Challenges Faced by Underrepresented Minority Students

Many underrepresented minority students enter college with a lower high school GPA, fewer math and science high school courses, and more concerns about paying for their education than non-underrepresented students.[10] As a consequence, they may feel behind when compared to their peers.

Underrepresented minorities also spend more hours working during high school and are more likely to work full-time during college, thereby taking away time they could spend on academics or extracurricular activities.[11]

Other challenges faced by these students may include a lack of role models that they can relate to and a curriculum that is more focused on mainstream students. Taken together, these stumbling blocks can cause great frustration and difficulty.

Challenges Faced by First-Generation College Students and Low-Income Students

According to Education Trust, 50.7 percent of students receiving a Pell Grant—a need-based grant for low-income undergraduates—gained their bachelor's degree in six years, compared to 56.4 percent of the general population and 65.9 students not receiving a Pell Grant at all.[12]

In many cases, first-generation and low-income students are one and the same. According to Dr. Mya Poe, coauthor of *Learning to Communicate in Science and Engineering: Case Studies from MIT,* "Many students from lower economic classes are often the first in their families to attend college and may not understand the hidden code of how to make their way through college," thus leading to difficulties in taking advantage of college resources. She continues, "Oftentimes, first generation or socioeconomically disadvantaged students are afraid to ask for help or clarification. They may not realize that they can go to the professor or TA for assistance or worry that their struggles are unique to them."

First generation and low-income students may also find it difficult to share some of their frustrations with their peers. According to Kaya, a student at Dartmouth College whom we interviewed: "When you are a low-income student, you may find yourself struggling with financial aid, and you don't know whether your friends will be able to understand your problems. You can't pay for dinners outside of the dining hall or pay for formals. You're already struggling to navigate college and social life during first year, but if you are a low-income student, you may also have to pick up work-study to pay for your education, and if you don't know how to balance everything, then you can struggle in your classes or forget to take care of yourself."

How to Succeed as a Student Underrepresented in STEM

Despite the aforementioned challenges, there are concrete ways to optimize your experience in college STEM classes.

1) Look for the Resources That Will Help You

Be proactive in identifying and taking advantage of resources that can help you to succeed in college. This may mean looking for sources of grants and scholarships designated for your underrepresented group, and utilizing tutoring programs if you feel like you might benefit from them. Understand that the administrators at your school—from the academic deans to the career counselors—are all there to help students navigate their way through academics and the real world. Outside of universities, you will never find such a comprehensive collection of professionals who are invested in helping you to succeed. Make use of them.

2) Talk to Faculty Members

Don't be afraid to reach out to faculty members. According to Dr. Lee Witters, former advisor to health professions students at Dartmouth College, the student should take the lead "because many faculty members are often busy and have many other responsibilities, you can't count on the faculty member to reach out to the student. The student has to take the initiative to initiate the conversation, to maintain that contact, and to recognize that this is a part of their education." Get to know your teachers, get their help when you are struggling and when you are succeeding, and get them personally invested in your success.

3) Believe That Your Ability Will Grow

You can resist the effects of stereotype threat by simply knowing about it and by maintaining a positive outlook on learning. And now that you know, you are off to a good start. Additionally, thinking of your abilities as capable of improvement instead of fixed and unchanging can have a significantly positive effect on grades and test scores.[13] Students who tell themselves that they are "not good at physics" are at a disadvantage compared to those students who think that studying and practice make people good at physics. Even if you encounter academic setbacks early in your college career, don't be deterred. Continue to believe in your capacity to improve.

4) Find a Community

Put yourself in a position where you can feel that you belong. According to Dr. Poe, "Underrepresented students who are in a learning community are much more likely to succeed on college campuses." Many students have doubts about their academic abilities from time to time, but positive reinforcement from others can help you regain confidence in yourself. Join professional, social, or community groups that will put you in contact with individuals who can encourage you and provide you with support and role models in the form of professors, administrators, upperclassmen, and fellow students.

Outside of your university are a number of professional organizations that exist to empower scientists and students from

Table A.1. Professional Organizations That Empower Scientists and Students from Underrepresented Communities

Underrepresented Demographic	Organizations/Professional Societies/Conferences
Female	Association for Women in Mathematics Association for Women in Science Center for Women in Science and Engineering Society of Women Engineers Women in Engineering Programs & Advocates Network
Black	National Organization for the Professional Advancement of Black Chemists and Chemical Engineers National Society of Black Engineers
Latino/Hispanic	MAES: A Society of Latino Engineers and Scientists The National Society of Hispanic Physicists Society of Hispanic Professional Engineers
Native American	American Indian Science and Engineering Society Society for Advancement of Chicanos and Native Americans in Science
First-Generation	I'm First (http://www.imfirst.org/)
General	National Action Council for Minorities in Engineering, Inc. Annual Biomedical Research Conference for Minority Students

underrepresented communities. Many of these organizations offer science meetings and programs to connect students with others like themselves.

5) Find a Mentor Who Can Support and Encourage You

According to an article in the *Economics of Education Review* authored by Wake Forest University economics professor Dr. Amanda Griffith, underrepresented students are more likely to continue in their STEM major if they seek out mentors who are also from underrepresented groups.[14] Dr. Griffith also notes that "female and minority students attending universities with a larger percentage of STEM PhD students that are women or minorities are more likely to persist in a STEM major."[15]

Find a sense of belonging by identifying a mentor, someone has gone through what you hope to go through in the future. The hardest thing about finding mentors is the process of first reaching out to them. Kaya, a student from Dartmouth College, advises that the student take the first step: "If you have someone you admire, [reaching out to them] may initially be awkward, but it could be as simple as saying 'Hi, I really look up to you, and I want to do what you do, and I'm hoping that we can have a conversation about your work and your advice on what I should do in the future.' Reach out to them whether they are in industry, whether they are students, professors, or famous people. You never know what will happen. And once you find a mentor, make sure to keep up the relationship. If they are busy people, they may forget to continually contact you. Reach out every month and get to know them. Contribute as a mentee by keeping up the relationship."

Conclusion

Don't let preconceived biases prevent you from performing at your best in college. There are many programs and opportunities that have specifically been created for women, underrepresented minority students, first-generation, and low-income college students. Seek out the resources at your institution to start off on the right foot, do not be afraid to seek guidance from administrators and your professors, and, above all, approach your courses with confidence. Find a mentor whom you like and who can open doors for you. You are already off to a great start.

Students Say: What Advice Would You Give to an Underrepresented STEM Student?

The most important thing an underrepresented student can do to ease the journey is to find a good mentor. I have been so fortunate to have several African American mentors to guide me through the process of applying to graduate school and navigating research environments. Because they had gone through the process themselves, they knew first-hand how to succeed in these environments, which may have very few underrepresented students. Building a community of students from similar backgrounds is also helpful because it facilitates a space where students can share stories, give tips, and ultimately help each other through this arduous process.

Anthony, Morehouse University

The road ahead will not be easy. College will present a new set of challenges that will feel especially difficult to you, and many of your private school peers will be having a much easier time. But, remember that the scientific field is one of the most exciting and rewarding career paths that you can take. What could be better than to be working to answer life's greatest questions? To invent something to make life easier for somebody else? All while working your brain at unbelievable capacity. If you step into college and you find yourself the last person in your lab or with a below-average score, just remember how much you've already gone through. Remember that the same resilience that got you through high school despite all your obstacles can be applied to academics. Remember that you are valuable, and your presence and perspective are assets to any institution of higher learning. In fact, we need people like you to pave the way for others. Now, that may seem like a lot of unfair pressure, but that was very important for me to remember in my hardest days.

Paloma, Williams College

Notes

Chapter One

1. Carl Sagan, "In Praise of Science and Technology," *New Republic* 176, no. 4 (1977): 21.
2. Christopher Drew, "Why Science Majors Change Their Minds (It's Just So Darn Hard)," *New York Times*, November 4, 2011.
3. US Congress Joint Economic Committee, "Stem Education: Preparing for the Jobs of the Future," website of US Congress Joint Economic Committee, http://www.jec.senate.gov/public, published 2012.
4. David Langdon, George McKittrick, David Beede, Beethika Khan, and Mark Doms, "STEM: Good Jobs Now and for the Future," website of US Department of Commerce Economics and Statistics Administration, http://www.esa.doc.gov/sites/default/files/stemfinalyjuly14_1.pdf, published 2011.
5. Ibid.
6. Anthony Carnevale, Nicole Smith, and Michelle Melton, "STEM: Science, Technology, Engineering, Mathematics," Georgetown University Center on Education and the Workforce website, https://cew.george town.edu/wp-content/uploads/2014/11/stem-complete.pdf.
7. David Langdon, George McKittrick, David Beede, Beethika Khan, and Mark Doms, "STEM: Good Jobs Now and for the Future," website of US Department of Commerce Economics and Statistics Administration, http://www.esa.doc.gov/sites/default/files/stemfinalyjuly14_1.pdf, published 2011.

8. SpencerStuart, "Leading CEOs: A Statistical Snapshot of S&P 500 Leaders," SpencerStuart website, https://www.spencerstuart.com, 2006.

9. Millstone Township Foundation for Educational Excellence, "STEM Education Is a Critical Component in Today's Society," Millstone Township website, http://www.mtfee.org, published 2013.

10. US Department of Education, "Preparing Our Children for the Future: Science, Technology, Engineering and Mathematics (STEM) Education in the 2011 Budget," website of the Executive Office of the President, http://www.whitehouse.gov/sites/default/files/stem_11_final.pdf, published 2010.

11. Brian Solomon, "Grading Disparity between Departments: Even Worse than You Thought," in *Dartblog*, 2012.

12. Ben Ost, "The Role of Peers and Grades in Determining Major Persistence in the Sciences," http://tigger.uic.edu/~bost/persist_science.pdf, published 2010.

13. Jenna Goudreau, "The 10 Worst College Majors," *Forbes*, October 11, 2012.

Chapter Two

1. E. Ophir, C. Nass, and A. D. Wagner, "Cognitive Control in Media Multitaskers," *Proceedings of the National Academy of Sciences* 106, no. 37 (2009): 15583–87. doi:10.1073/pnas.0903620106.

2. FOMO, Oxford Dictionaries website, Oxford University Press, http://www.oxforddictionaries.com/us/definition/american_english/FOMO.

3. Michelle Voss, Lindsay Nagamatsu, Teresa Liu-Ambrose, and Arthur Kramer, "Exercise, Brain and Cognition across the Lifespan," *Journal of Applied Physiology* 10, no. 1152 (2011).

4. "How Much Physical Activity Is Needed?" US Department of Agriculture website, ChooseMyPlate.gov.

5. Michelle Florence, Mark Asbridge, and Paul Veugelers, "Diet Quality and Academic Performance," *Journal of School Health* 78, no. 4 (2008): 209–15.

6. Henry Wechsler and Toben F. Nelson, "What We Have Learned from the Harvard School of Public Health College Alcohol Study: Focusing Attention on College Student Alcohol Consumption and the Environ-

mental Conditions That Promote It," *Journal of Studies on Alcohol and Drugs* 69, no. 4 (2008): 481–90.

7. "How Much Sleep Do I Need?" Centers for Disease Control and Prevention website, http://www.cdc.gov, published July 1, 2013.

8. Ana Gomes, Jose Tavares, and Maria Azevedo, "Sleep and Academic Performance in Undergraduates: A Multi-measure, Multi-predictor Approach," *Chronobiology International* 28, no. 9 (2011): 786–801.

9. Rusty Lindquist, "Sleep Deprivation: How You Sabotage Your Own Success," Life Engineering Blog, February 21, 2011, http://life-engineering .com/sleep-deprivation-like-being-legally-drunk.

10. National Institute of Mental Health, "Mental Illness Exacts Heavy Toll, Beginning in Youth," National Institute of Mental Health website, http://www.nimh.nih.gov, published 2005.

11. Ibid.

12. American College Health Association (ACHA), "National College Health Assessment: Reference Group Executive Summary, Spring 2012," ACHA website, http://www.acha-ncha.org, published 2012.

Chapter Three

1. Lionel Giles, *Sun Tzu on the Art of War* (London: Luzac, 1910).

2. "Stock Up on Ramen: Average Cost of College Rises Again," *USA Today College*, November 13, 2014.

3. Walter Pauk, *How to Study in College*, 7th ed. (Boston: Houghton Mifflin, 2001).

4. Hedwig Von Restorff, "Über die wirkung von bereichsbildungen im spurenfeld," *Psychologische Forschung* 18, no. 1 (1933): 299–342.

5. John Dunlosky, Katherine Rawson, Elizabeth Marsh, Mitchell Nathan, and Daniel Willingham, "Improving Students' Learning with Effective Learning Techniques: Promising Directions from Cognitive and Educational Psychology," *Psychological Science in the Public Interest* 14, no. 1 (2013): 4–58.

6. Kathy Slobogin, "Survey: Many Students Say Cheating's OK," CNN.com, April 5, 2002.

7. Cal Newport, "Deep Habits: Should You Track Hours or Milestones?" Study Hacks Blog: Decoding the Patterns of Success, March 23, 2014.

8. Ibid.

9. Charles A. Morgan, Ann M. Rasmusson, Sheila Wang, Gary Hoyt, Richard L. Hauger, and Gary Hazlett, "Neuropeptide-Y, Cortisol, and Subjective Distress in Humans Exposed to Acute Stress: Replication and Extension of Previous Report," *Biological Psychiatry* 52, no. 2 (2002): 136–42.

Chapter Four

1. The College Board, "The College Major: What It Is and How to Choose One," https://bigfuture.collegeboard.org/explore-careers/college-majors/the-college-major-what-it-is-and-how-to-choose-one, published 2015.

2. Unigo, "When Do I Need to Choose a Major?" *US News & World Report*, May 25, 2011, http://www.usnews.com/education/blogs/college-admissions-experts/2011/05/25/when-do-i-need-to-choose-major.

3. Lewis Thomas, "On Science and Uncertainty," *Discover* magazine, 1980, 59.

4. National Science Foundation, National Center for Science and Engineering Statistics, "Table 35. Employment Sector of Recent Graduates with Bachelor's Degrees in Science, Engineering, or Health, by Occupation: October 2010." http://ncsesdata.nsf.gov/recentgrads/2010/html/RCG2010_DST35.html.

5. Neta P. Fogg, Thomas F. Harrington, Paul E. Harrington, and Laurence Shatkin, *College Majors Handbook* (St. Paul: JIST, 2012).

6. Sasha Gurke, "Why Engineers Could Make the Best Business Leaders," *Business Insider*, http://www.businessinsider.com/why-engineers-make-the-best-business-leaders-2011–12#ixzz3C246v8Dn.

7. Albert Einstein, "Professor Einstein Writes in Appreciation of a Fellow-Mathematician," *New York Times*, http://www-history.mcs.st-and.ac.uk/Obits2/Noether_Emmy_Einstein.html.

8. G. E. Pugh, *The Biological Origin of Human Values* (London: Routledge & Kegan Paul, 1978).

9. Society for Neuroscience, "NeuroJobs Career Center," Society for Neuroscience website, http://www.sfn.org/careers-and-training/neurojobs-career-center/careers-in-neuroscience.

Chapter Five

1. *Nature*: "For Authors: Getting Published in *Nature*: The Editorial Process," http://www.nature.com/nature/authors/get_published/; *Science*: "The *Science* Contributors FAQ," http://www.sciencemag.org/site /feature/contribinfo/faq/#pct_faq.

2. Alec Wilkinson, "The Pursuit of Beauty," *New Yorker*, February 2, 2015.

3. NIH Office of the Ombudsman, "The Ombudsman's Role at NIH," May 30, 2015.

4. National Science Foundation, "Grant Policy Manual," http://www.nsf. gov/pubs/2002/nsf02151/gpm2.jsp#210.

5. "Summer Internship Program FAQs," Summer Internship Program at the NIH, https://www.training.nih.gov/resources/faqs/summer _interns#q17.

6. "How to Read a Scientific Paper," Science Buddies, http://www.science buddies.org/science-fair-projects/top_science-fair_how_to_read_a _scientific_paper.shtml, accessed January 5, 2016.

7. Ibid.

8. Phil Dee, "Writing Papers and Abstracts," in *Building a Successful Career in Scientific Research: A Guide for PhD Students and Postdocs* (Cambridge: Cambridge University Press, 2006), 33.

9. Ibid., 34.

10. Ibid.

11. Ibid.

12. Phil Dee, "Conferences and Poster Presentations," in *Building a Successful Career in Scientific Research: A Guide for PhD Students and Postdocs* (Cambridge: Cambridge University Press, 2006), 48.

13. Yoo Jung Kim, "Why Every Science Student Should Attend a Conference—The Student Blog," The Student Blog. February 24, 2014.

Chapter Six

1. "Jobvite 2014 Social Recruiting Survey," October 16, 2014.

2. "Frequently Asked Questions," The Rhodes Trust.

Chapter Seven

1. "Earnings and Unemployment Rates by Educational Attainment," US Department of Labor Bureau of Labor Statistics, http://www.bls.gov /emp/ep_chart_001.htm.

2. "Academic Graduate Study: Master's and PhD Programs," Career Services Center, University of California, San Diego, http://career.ucsd .edu/_files/Academic_Grad_Study_Guide.

3. "Project Information," Council of Graduate Schools PhD Completion Project, http://www.phdcompletion.org/information/index.asp.

4. Sheila Nataraj Kirby, Robert Sowell, Nathan Bell, and Scott Naftel, "PhD Completion: Findings from the Exit Surveys," Council of Graduate Schools, 2009.

5. Association of American Medical Colleges, "Table 18: MCAT and GPA for Applicants and Matriculants to US Medical Schools by Primary Undergraduate Major, 2013."

6. Association of American Medical Colleges, ed., *AAMC Resources for Pre-Med Students* (Washington, DC: Association of American Medical Colleges, 2014).

7. Association of American Medical Colleges, "Table 24: MCAT and GPA Grid for Applicants and Acceptees to US Medical Schools, 2011–2013 (Aggregated)."

8. Association of American Medical Colleges, "Table 17: MCAT Scores and GPAs for Applicants and Matriculants to US Medical Schools, 2002–2013."

9. Association of American Medical Colleges, "Table 6: Age of Applicants to US Medical Schools at Anticipated Matriculation by Sex and Race and Ethnicity, 2009–2012."

10. NGS Movement, "Alternatives to Teach for America," http://ngsmove ment.org/2014/01/19/alternatives-to-teach-for-america/.

11. Shawn O'Connor, "In Law School Admissions, STEM Sells," in *Law Admissions Lowdown: US News & World Report*, 2012.

12. Carol Leach, "Selecting a Major for Law School," Health Professions and Pre-Law Center at Indiana University Bloomington, 2007.

13. "About the LSAT," Law School Admission Council, http://www.lsac.org /jd/lsat/about-the-lsat/.

14. "LSAT Score Conversion," Alphascore, http://www.alphascore.com /resources/lsat-score-conversion/.

15. David Lat, "Nationwide Layoff Watch: Latham Cuts 440," in *Above the Law*, 2009.

16. Jonathan Berr, "Shrinking Law Schools Face Financial Devastation," *Fiscal Times*, March 13, 2014.

17. Alaina Levin, "Wanted: BS and MS Scientists in Life Sciences Industries," *Science*, January 13, 2012, http://sciencecareers.sciencemag.org /career_magazine/previous_issues/articles/2012_01_13/science.opms .r1200113.

18. "Pfizer Jobs," Pfizer, 2014.

19. "Create Your Business Plan," US Small Business Administration.

20. "Admissions Class Profile," Harvard Business School, http://www.hbs .edu/mba/admissions/class-profile/Pages/default.aspx.

21. Diana Middleton, "The Top MBA Programs If You're in a Hurry," *Wall Street Journal*, September 16, 2009.

22. "Admissions: College Students 2 + 2," Harvard Business School, http:// www.hbs.edu/mba/admissions/application-process/Pages/2-plus-2 -application-process.aspx.

Chapter Eight

1. Paul C. Lauterbur, "All Science Is Interdisciplinary: From Magnetic Moments to Molecules to Men," *Bioscience Reports* 24, no. 3 (2004): 165–78.

2. Yoo Jung Kim, "Kim: Breaking Boundaries," *Dartmouth*, March 6, 2014, Opinion sec. http://www.thedartmouth.com/2014/03/06/kim-breaking -boundaries/.

3. Rebecca Davis, "The Doctor Who Championed Hand-Washing and Briefly Saved Lives," NPR, January 12, 2015.

4. Yoo Jung Kim, "How Undergraduate Journals Foster Scientific Communication: Sci-Ed," Public Library of Science, Sci-Ed Blog, January 12, 2015.

5. Cal Newport, "16: Always Be Working on a Grand Project," in *How to Win at College: Simple Rules for Success from Star Students* (New York: Broadway, 2005), 41.

6. Yoo Jung Kim, "Kim: Embracing Disappointments," *Dartmouth*, September 4, 2012, Opinion sec. http://thedartmouth.com/2012/09/04/kim-embracing-disappointments/.

Appendix

1. "Science and Engineering Indicators 2014: Chapter 3. Science and Engineering Labor Force," National Science Foundation.
2. Ibid.
3. Michele Paludi and William Bauer, "Goldberg Revisited: What's in an Author's Name," *Sex Roles* 9, no. 3 (1983): 387–90.
4. Claude M. Steele and Joshua Aronson, "Stereotype Threat and the Intellectual Test Performance of African Americans," *Journal of Personality and Social Psychology* 69, no. 5 (1995): 797.
5. Daniel Hemel, "Summers' Comments on Women and Science Draw Ire," *Harvard Crimson*, January 14, 2005, http://www.thecrimson.com/article/2005/1/14/summers-comments-on-women-and-science/.
6. J. S. Hyde, S. M. Lindberg, M. C. Linn, A. B. Ellis, and C. C. Williams, "Diversity: Gender Similarities Characterize Math Performance," *Science* 321, no. 5888 (2008): 494–95.
7. Lee Jussim and Jacquelynne Eccles, "Teacher Expectations II: Construction and Reflection of Student Achievement," *Journal of Personality and Social Psychology* 63, no. 6 (1992): 947–61.
8. You-Jung Kim, "Gender Discrimination in Science," *Korea Times*, December 10, 2008.
9. C. Hill, C. Corbett, and A. St. Rose, *Why So Few? Women in Science, Technology, Engineering, and Mathematics* (Washington, DC: AAUW, 2010).
10. Sylvia Hurtado, Christopher B. Newman, Minh C. Tran, and Mitchell J. Chang, "Improving the Rate of Success for Underrepresented Racial Minorities in STEM Fields: Insights from a National Project," *New Directions for Institutional Research*, no. 148 (2010).
11. Ibid.
12. The Education Trust, "The Pell Partnership: Ensuring a Shared Responsibility for Low-Income Student Success," September 1, 2015.
13. M. Johns, T. Schmader, and A. Martens, "Knowing Is Half the Battle: Teaching Stereotype Threat as a Means of Improving Women's Math

Performance," *Psychological Science* 16, no. 3 (2005): 175–79; Joshua Aronson, Carrie B. Fried, and Catherine Good, "Reducing the Effects of Stereotype Threat on African American College Students by Shaping Theories of Intelligence," *Journal of Experimental Social Psychology* 38, no. 2 (2002): 113–25.

14. Amanda L. Griffith, "Persistence of Women and Minorities in STEM Field Majors: Is It the School That Matters?" *Economics of Education Review* 29, no. 6 (2010): 911–22.

15. Ibid.

Index